TWENTY FIRST CENTURY SCIENCE

Project Directors
Angela Hall   Emma Palmer
Robin Millar   Mary Whitehouse

Editors
Angela Hall   Emma Palmer
Mary Whitehouse

Authors

Cris Edgell   John Lazonby   Emily Perry   Carol Tear
Mike Kalvis   Ted Lister   Cliff Porter   Vicky Wong
                 Robin Millar   Mike Shipton

THE UNIVERSITY of York
THE SALTERS' INSTITUTE

Nuffield Foundation

OCR RECOGNISING ACHIEVEMENT   OXFORD UNIVERSITY PRESS
Official Publisher Partnership

Contents

How to use this book 4	Making sense of graphs 9
Structure of assessment 6	Controlled assessment 12
Command words 8	

B4 The processes of life 14

A	Features of all living things	16	G	Minerals from the soil 28
B	Enzymes	18	H	The rate of photosynthesis 29
C	Keeping the best conditions for enzymes	20	I	Environments and adaptations 32
			J	Energy for life 34
D	Enzymes at work in plants	22	K	Energy without oxygen 36
E	Diffusion and gas exchange in plants	24	L	Useful products from respiration 38
F	Osmosis	26		Summary 40

C4 Chemical patterns 44

A	The periodic table	46	H	Electrons and the periodic table 60
B	The alkali metals	48	I	Salts 62
C	Chemical equations	50	J	Ionic theory 64
D	The halogens	52	K	Ionic theory and atomic structure 66
E	The discovery of helium	54	L	Chemical species 68
F	Atomic structure	56		Summary 70
G	Electrons in atoms	58		

P4 Explaining motion 74

A	Forces and interactions	76	G	Forces and motion 88
B	Getting moving	78	H	Car safety 91
C	Friction	80	I	Laws of motion 94
D	Vertical forces	82	J	Work and energy 96
E	Describing motion	84		Summary 100
F	Picturing motion	86		

B5 Growth and development 104

A	Growing and changing	106	G	The mystery of inheritance 120
B	Growing plants	110	H	Making proteins 122
C	Phototropism	112	I	Specialised cells – special proteins 124
D	A look inside the nucleus	114	J	Stem cells 128
E	Making new cells	116		Summary 130
F	Sexual reproduction	118		

C5 Chemicals of the natural environment — 134

A Chemicals in spheres	136	F Carbon minerals – hard and soft	148
B Chemicals of the atmosphere	138	G Metals from the lithosphere	150
C Chemicals of the hydrosphere	140	H Structure and bonding in metals	156
D Detecting ions in salts	142	I The life cycle of metals	158
E Chemicals of the lithosphere	146	Summary	160

P5 Electric circuits — 164

A Electric charge	166	F Electrical power	182
B Electric currents in circuits	169	G Magnets and motors	184
C Branching circuits	172	H Generating electricity	186
D Controlling the current	173	I Distributing electricity	188
E Potential difference	178	Summary	190

B6 Brain and mind — 194

A What is behaviour?	196	F Learnt behaviour	208
B Simple reflexes in humans	198	G Human learning	212
C Your nervous system	202	H What is memory?	214
D Synapses	204	Summary	220
E The brain	206		

C6 Chemical synthesis — 224

A The chemical industry	226	F Rates of reaction	238
B Acids and alkalis	228	G Stages in chemical synthesis	244
C Salts from acids	232	H Chemical quantities	248
D Purity of chemicals	234	Summary	250
E Energy changes in chemical reactions	236		

P6 Radioactive materials — 254

A Radioactive materials	256	G Half-life	270
B Atoms and nuclei	258	H Medical imaging and treatment	272
C Inside the atom	260	I Nuclear power	274
D Using radioactive isotopes	262	J Nuclear waste	276
E Radiation all around	264	K Nuclear fusion	278
F Living with radon	267	Summary	280

Glossary	284	**Appendices**	293
Index	290		

How to use this book

Welcome to Twenty First Century Science. This book has been specially written by a partnership between OCR, The University of York Science Education Group, The Nuffield Foundation Curriculum Programme, and Oxford University Press.

On these two pages you can see the types of page you will find in this book, and the features on them. Everything in the book is designed to provide you with the support you need to help you prepare for your examinations and achieve your best.

Module Openers

Why study?: This explains why what you are about to learn is useful to scientists.

Find out about: Every module starts with a short list of the things you'll be covering.

Ideas about Science: Here you can read about the key ideas about science covered in this module.

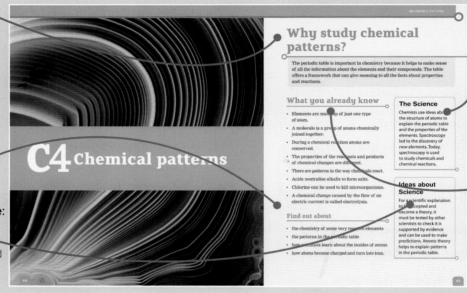

The Science: This box summarises the science behind the module you're about to study.

What you already know: This list is a summary of the things you've already learnt that will come up again in this module. Check through them in advance and see if there is anything that you need to recap on before you get started.

Main Pages

Find out about: For every part of the book you can see a list of the key points explored in that section.

Summary box: This box sums up the main ideas covered on these pages.

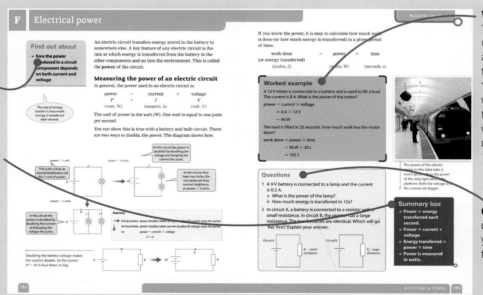

Worked examples: These help you understand how to use an equation or to work through a calculation. You can check back whenever you use the calculation in your work to make sure you understand it.

Questions: Use these questions to see if you've understood the topic.

You should know: This is a summary of the main ideas in the unit. You can use it as a starting point for revision, to check that you know about the big ideas covered.

Visual summary: Another way to start revision is to use a visual summary, linking ideas together in groups so that you can see how one topic relates to another. You can use this page as a starting point for your own summary.

Science Explanations

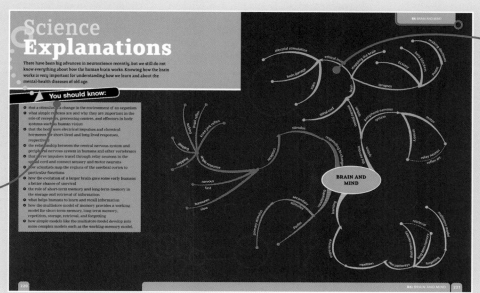

Ideas about Science: For every module this page summarises the ideas about science that you need to understand.

Review Questions: You can begin to prepare for your exams by using these questions to test how well you know the topics in this module.

Ideas about Science and Review Questions

Structure of assessment

Matching your course
What's in each module?
As you go through the book you should use the module opener pages to understand what you will be learning and why it is important. The table below gives an overview of the main topics each module includes.

B4
- How do chemical reactions take place in living things?
- How do plants make food?
- How do living organisms obtain energy?

C4
- What are the patterns in the properties of elements?
- How do chemists explain the patterns in the properties of elements?
- How do chemists explain the properties of Group 1 and Group 7 elements?

P4
- How can we describe motion?
- What are forces?
- What is the connection between forces and motion?
- How can we describe motion in terms of energy changes?

B5
- How do organisms develop?
- How does an organism produce new cells?
- How do genes control growth and development within the cell?

C5
- What types of chemicals make up the atmosphere?
- What reactions happen in the hydrosphere?
- What types of chemicals make up the lithosphere?
- How can we extract useful metals from minerals?

P5
- Electric current – a flow of what?
- What determines the size of current in an electric circuit and the energy it transfers?
- How do series and parallel circuits work?
- How is mains electricity produced? How are voltages and currents induced?
- How do electric motors work?

B6
- How do organisms respond to changes in their environment?
- How is information passed through the nervous system?
- What can we learn through conditioning?
- How do humans develop complex behaviour?

C6
- Chemicals and why we need them.
- Planning, carrying out, and controlling a chemical synthesis.

P6
- Why are some materials radioactive?
- How can radioactive materials be used and handled safely, including wastes?

How do the modules fit together?
The modules in this book have been written to match the specification for GCSE Additional Science. In the diagram to the right you can see that the modules can also be used to study parts of GCSE Biology, GCSE Chemistry, and GCSE Physics courses.

	GCSE Biology	GCSE Chemistry	GCSE Physics
GCSE Science	B1	C1	P1
	B2	C2	P2
	B3	C3	P3
GCSE Additional Science	B4	C4	P4
	B5	C5	P5
	B6	C6	P6
	B7	C7	P7

GCSE Additional Science assessment

The content in the modules of this book matches the modules of the specification.

Twenty First Century Science offers two routes to the GCSE Additional Science qualification, which includes different exam papers depending on the route you take.

The diagrams below show you which modules are included in each exam paper. They also show you how much of your final mark you will be working towards in each paper.

	Unit	Modules Tested			Percentage	Type	Time	Marks Available
Route 1	A162	B4	B5	B6	25%	Written Exam	1 h	60
Route 1	A172	C4	C5	C6	25%	Written Exam	1 h	60
Route 1	A182	P4	P5	P6	25%	Written Exam	1 h	60
Route 1	A154	Controlled Assessment			25%		4.5–6 h	64
Route 2	A151	B4	C4	P4	25%	Written Exam	1 h	60
Route 2	A152	B5	C5	P5	25%	Written Exam	1 h	60
Route 2	A153	B6	C6	P6	25%	Written Exam	1 h	60
Route 2	A154	Controlled Assessment			25%		4.5–6 h	64

Command words

The list below explains some of the common words you will see used in exam questions.

Calculate
Work out a number. You can use your calculator to help you. You may need to use an equation. The question will say if your working must be shown. (Hint: don't confuse with 'Estimate' or 'Predict'.)

Compare
Write about the similarities and differences between two things.

Describe
Write a detailed answer that covers what happens, when it happens, and where it happens. Talk about facts and characteristics. (Hint: don't confuse with 'Explain'.)

Discuss
Write about the issues related to a topic. You may need to talk about the opposing sides of a debate, and you may need to show the difference between ideas, opinions, and facts.

Estimate
Suggest an approximate (rough) value, without performing a full calculation or an accurate measurement. Don't just guess – use your knowledge of science to suggest a realistic value. (Hint: don't confuse with 'Calculate' and 'Predict'.)

Explain
Write a detailed answer that covers how and why a thing happens. Talk about mechanisms and reasons. (Hint: don't confuse with 'Describe'.)

Evaluate
You will be given some facts, data, or other kind of information. Write about the data or facts and provide your own conclusion or opinion on them.

Justify
Give some evidence or write down an explanation to tell the examiner why you gave an answer.

Outline
Give only the key facts of the topic. You may need to set out the steps of a procedure or process – make sure you write down the steps in the correct order.

Predict
Look at some data and suggest a realistic value or outcome. You may use a calculation to help. Don't guess – look at trends in the data and use your knowledge of science. (Hint: don't confuse with 'Calculate' or 'Estimate'.)

Show
Write down the details, steps, or calculations needed to prove an answer that you have given.

Suggest
Think about what you've learnt and apply it to a new situation or context. Use what you have learnt to suggest sensible answers to the question.

Write down
Give a short answer, without a supporting argument.

Top Tips

Always read exam questions carefully, even if you recognise the word used. Look at the information in the question and the number of answer lines to see how much detail the examiner is looking for.

You can use bullet points or a diagram if it helps your answer.

If a number needs units you should include them, unless the units are already given on the answer line.

Making sense of graphs

Scientists use graphs and charts to present data clearly and to look for patterns in the data. You will need to plot graphs or draw charts to present data in the practical investigation and then describe and explain what the data is showing. Examination questions may also ask you to describe and explain what a graph is telling you.

Describing the relationship between variables

The pattern of points plotted on a graph can show whether two **factors** are related.

To describe the relationship in detail you should:
- read the axes and check the units used
- identify distinct phases of the graph – where the gradient is different
- use data when describing the changes.

Gradient of the graph

The **gradient** of the graph describes the way one variable changes relative to the other. Often the x axis is the time axis. The gradient then describes the rate of change.

Look at this graph, which shows the product of a chemical reaction being produced over time. The gradient of the graph gives the rate of the chemical reaction.

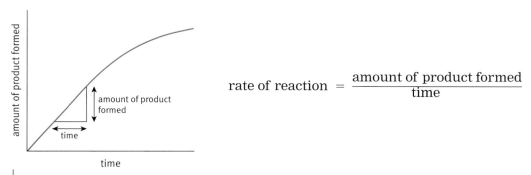

$$\text{rate of reaction} = \frac{\text{amount of product formed}}{\text{time}}$$

Graph showing how rate is calculated for a chemical reaction.

Look at this graph, which shows the distant travelled by a car over a period of time. The gradient of the graph gives the speed of the car.

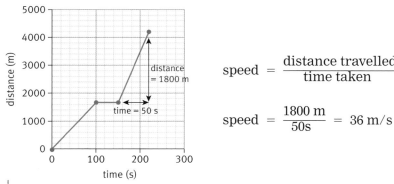

$$\text{speed} = \frac{\text{distance travelled}}{\text{time taken}}$$

$$\text{speed} = \frac{1800 \text{ m}}{50 \text{s}} = 36 \text{ m/s}$$

Calculating the gradient of a distance–time graph to find the speed.

Calculating reacting masses and percentage yields

Problem *How much carbon do you need to react with 8.1 g of zinc oxide?*

Start with the balanced symbol equation:

zinc oxide + carbon ⟶ zinc + carbon monoxide
ZnO + C ⟶ Zn + CO

Find the relative atomic mass of each element in the equation on a periodic table. Work out the relative formula masses by adding up the relative atomic masses of all the atoms in each formula.

Reactants: Relative formula mass of ZnO = 65 + 16 = 81
Relative formula mass of C = 12

Products: Relative formula mass of Zn = 65
Relative formula mass of CO = 12 + 16 = 28

Write the relative formula masses under each chemical in the balanced equation and convert to reacting masses by including units. These can be grams, kilograms, or tonnes, depending on the data in the question. The units must be the same for each of the values.

ZnO + C ⟶ Zn + CO
81 g   12 g   65 g   28 g

So 12 g of carbon (C) reacts with 81 g of zinc oxide (ZnO) to make 65 g of zinc (Zn) and 28 g of carbon monoxide (CO).

To find out how much carbon you need to react with 8.1 g of zinc oxide you need to scale the reacting masses to the amounts actually used.

If we started with one tenth of the amount of zinc oxide (8.1 g), it would react with one tenth of the amount of carbon (1.2 g).

Another way of scaling the reacting masses to the amounts actually used is to use a mathematical formula like this one.

$$\text{mass of carbon (C)} = \text{mass of zinc oxide (ZnO)} \times \frac{\text{relative formula mass of carbon (C)}}{\text{relative formula mass of zinc oxide (ZnO)}}$$

$$\text{mass of carbon (C)} = 8.1 \text{ g} \times \frac{12}{81}$$

$$= 1.2 \text{ g}$$

Problem *What is the percentage yield of zinc for the same reaction, if the theoretical yield is 6.5 g of zinc, and the actual yield is 5.4 g of zinc?*

To work out the theoretical yield use the mathematical formula:

$$\text{percentage yield} = \frac{\text{actual yield (g)}}{\text{theoretical yield (g)}} \times 100$$

Use the data given in the question to find the percentage yield:

actual yield = 5.4 g
theoretical yield = 6.5 g

$$\text{percentage yield} = \frac{5.4 \text{ g}}{6.5 \text{ g}} \times 100$$

$$= 83\%$$

Frequency data

Frequency graphs or charts show the number of times a data value occurs. For example, if four students have a pulse rate of 86, then the data value 86 has a frequency of four.

A large data set with lots of different values can be arranged into class intervals (or groups). Collecting data in class intervals can be done by tallying. It works well to have data arranged in five or six class intervals.

Class interval	Tally	Frequency
60–65	|	1
65–70	||||	4
70–75	⋕⋕ ⋕⋕ ||	12
75–80	⋕⋕ |||	8
80–85	⋕⋕	5
85–90	|	1
	Total	31

A data set of pulse rates from a class of 31 pupils tallied in a frequency table.

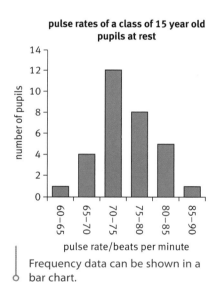

Frequency data can be shown in a bar chart.

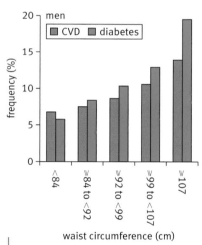

Sometimes frequency graphs have % as the units on the y axis. CVD: cardiovascular disease.

Q Write down a question that can be answered from each of these graphs.

Range and mean

Statistics are used to describe data. Useful statistics to describe the pulse rate data are the range and mean of the data.

The range of this data set can be expressed as 'between x beats per minute (lowest value) and y beats per minute (highest value)'

The mean is the total of all the values divided by the number of data points. It is an estimate of the true value of the variable being measured.

Q Write down two statements about the pulse rate data.

Q If you are comparing the pulse rates of two different classes in your school, why would it be useful to have both statistics (mean and range)?

Controlled assessment

In GCSE Additional Science the controlled assessment counts for 25% of your total grade. Marks are given for a practical investigation.

Your school or college may give you the mark schemes for this.

This will help you understand how to get the most credit for your work.

Practical investigation (25%)

Investigations are carried out by scientists to try and find the answers to scientific questions. The skills you learn from this work will help prepare you to study any science course after GCSE.

To succeed with any investigation you will need to:
- choose a question to explore
- select equipment and use it appropriately and safely
- design ways of making accurate and reliable observations
- relate your investigation to work by other people investigating the same ideas.

Your investigation report will be based on the data you collect from your own experiments. You will also use information from other people's research. This is called secondary data.

You will write a full report of your investigation. Marks will be awarded for the quality of your report. You should:
- make sure your report is laid out clearly in a sensible order
- use diagrams, tables, charts, and graphs to present information
- take care with your spelling, grammar, and punctuation, and use scientific terms where they are appropriate.

Marks will be awarded under five different headings.

Strategy
- Develop a hypothesis to investigate.
- Choose a procedure and equipment that will give you reliable data.
- Carry out a risk assessment to minimise the risks of your investigation.
- Describe your hypothesis and plan using correct scientific language.

Collecting data
- Carry out preliminary work to decide the range.
- Collect data across a wide enough range.
- Collect enough data and check its reliability.
- Control factors that might affect the results.

Analysis
- Present your data to make clear any patterns in the results.
- Use graphs or charts to indicate the spread of your data.
- Use appropriate calculations such as averages and gradients of graphs.

Evaluation
- Describe and explain how you could improve your method.
- Discuss how repeatable your evidence is, accounting for any outliers.

Review
- Comment, with reasons, on your confidence in the secondary data you have collected.
- Compare the results of your investigation to the secondary data.
- Suggest ways to increase the confidence in your conclusions.

Tip

The best advice is 'plan ahead'. Give your work the time it needs and work steadily and evenly over the time you are given. Your deadlines will come all too quickly, especially if you have coursework to do in other subjects.

Secondary data

Once you have collected the data from your investigation you should look for some secondary data relevant to your hypothesis. This will help you decide how well your data agrees with the findings of other scientists. Your teacher will give you secondary data provided by OCR, but you should look for further sources to help you evaluate the quality of all your data. Other sources of information could include:
- experimental results from other groups in your class or school
- text books
- the Internet.

When will you do this work?

Your school or college will decide when you do your practical investigation. If you do more than one investigation, they will choose the one with the best marks.

Your investigation will be done in class time over a series of lessons.

You may also do some research out of class.

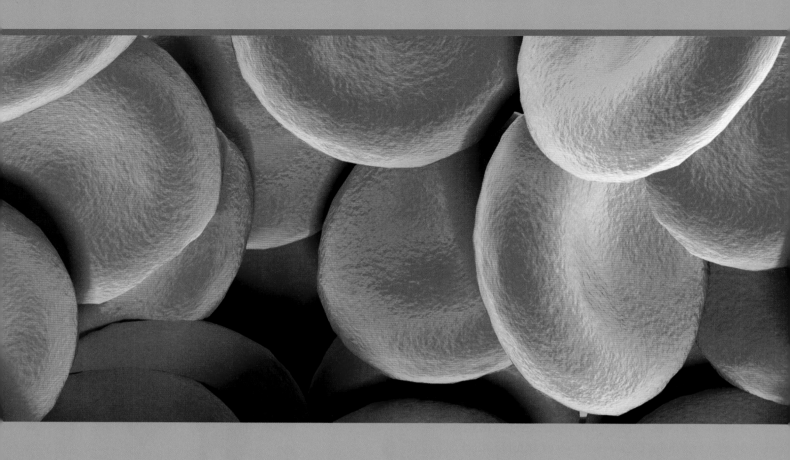

B4 The processes of life

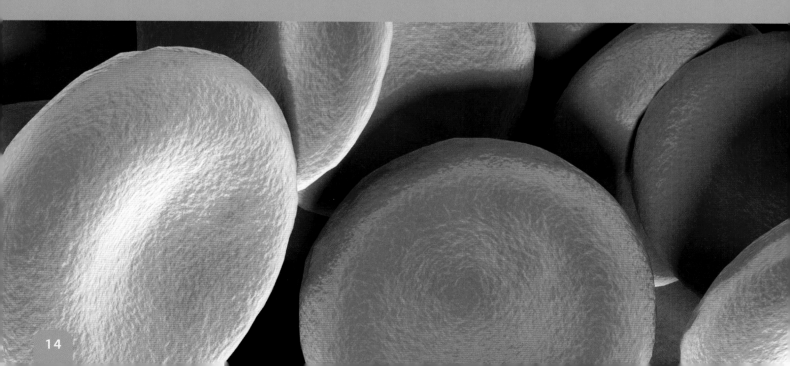

B4: THE PROCESSES OF LIFE

Why study the processes of life?

All living things need to be able to generate energy, repair damage, and grow new cells. Enzymes speed up these chemical reactions. They need specific conditions to work at their best. Understanding the processes of life helps us to treat diseases, improve food production, and produce sustainable fuels.

What you already know

- Instructions to control how an organism develops and functions are found in the nucleus of its cells and are called genes.
- Proteins may be functional, for example, enzymes such as amylase.
- Living organisms are adapted to their environment.
- Nearly all organisms are ultimately dependent on energy from the Sun.

Find out about

- the way that enzymes speed up chemical reactions in all living cells
- how microbes can be used to make useful products
- how investigations can explore the relationship between changing variables and outcomes
- the fact that photosynthesis captures energy from the Sun and respiration releases energy
- how different molecules move in and out of cells.

The Science

Photosynthesis captures energy from sunlight. The energy is stored when carbon dioxide and water are combined to make sugars. Respiration breaks down sugars and releases energy. Ultimately, the process of photosynthesis powers all life on Earth.

Ideas about Science

To investigate the way outcomes are affected when certain factors are changed, it is important to control any other factors that may also affect the outcome.

A | Features of all living things

> **Find out about**
> - the processes carried out by all living things

All living things are made up of **cells**. Some are just a single cell while others, like you, are made from billions of cells. You may not think you are like single-celled **bacteria** but there are some processes that all living things perform.

Move.

Bacteria and animals move to find food, escape predators, or find better conditions. Plants grow to find sunlight.

Respire.

Energy is needed to carry out cell processes. Respiration is a series of chemical reactions that releases energy from food.

Sense.

Living things sense and respond to their surroundings. Plants grow towards sunlight and woodlice run away from it.

Grow.

Bacteria grow and divide to form new bacteria. Plants and animals are made from billions of cells.

Excrete.

Living cells produce wastes. Wastes are removed by excretion. Carbon dioxide is a waste product of respiration.

Feed.

Living things get a supply of energy from their food. Plants make their own food during photosynthesis.

Reproduce.

All living things eventually die. Reproduction makes new generations.

Life processes

There is a biological chemical factory in every cell. It takes small molecules and builds them into large molecules. It uses chemical reactions to build itself, copy itself, and repair itself. It takes large food molecules and breaks these down to release energy in respiration.

All these reactions in living organisms are catalysed by **enzymes**. A catalyst speeds up a reaction but doesn't get used up.

Using life processes: making enzymes

Large tanks, like the ones in the picture, are used to grow bacteria. Their enzymes can be harvested. The tank contains a nutrient solution, and conditions are controlled for rapid growth.

As the bacteria grow, they release their enzymes. When the nutrients are used up, the contents are filtered to remove the bacteria. The enzymes are then purified. Enzymes are used in the food industry, textiles industry, and in the home as biological washing powders.

Bacteria are grown in this fermenter. They make enzymes that are used to make denim jeans look faded.

Like any living organism, bacteria need to take in food. They make and release enzymes into their surroundings. These break down any food molecules so that they can be absorbed into the microbe.

Every one of this plant's cells is a biological chemical factory.

Summary box
- All living things are made of cells.
- All living things:
  - move
  - respire
  - sense
  - grow
  - reproduce
  - excrete
  - feed
- Enzymes speed up the reactions in cells.

Questions

1. List the processes that all living things are able to do.
2. Name the process that generates energy for living cells.
3. What things can you sense in your surroundings? How do you respond? Give two examples.

B Enzymes

Find out about

- why you cannot live without enzymes
- how enzymes work

What are enzymes?

The chemical reactions that take place in cells are helped by enzymes. Enzymes are **catalysts** – they speed up chemical reactions in living things. They are **proteins** – large molecules made up of long chains of **amino acids**. The amino acid chains are different in each protein, so they fold up into different shapes. Genes in the nucleus of the cell carry the instructions for making the protein. An enzyme's shape is very important to how it works.

How do enzymes work?

Some enzymes break down large molecules into smaller ones. Others join small molecules together.

In all cases, the molecules must fit exactly into a part of the enzyme called the **active site**. It is a bit like fitting the right key in a lock. So scientists call the explanation of how an enzyme works the **lock-and-key model**.

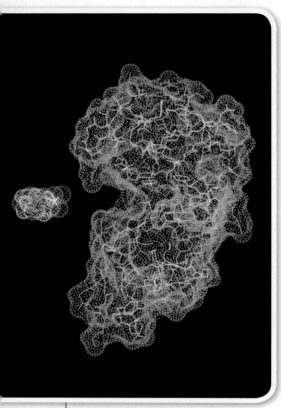

A computer graphic of an enzyme, its active site, and the product of a reaction.

The lock-and-key model of enzyme function. The diagram is simplified.

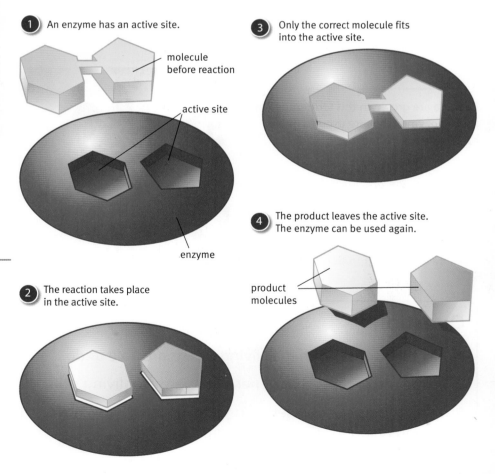

1. An enzyme has an active site.
2. The reaction takes place in the active site.
3. Only the correct molecule fits into the active site.
4. The product leaves the active site. The enzyme can be used again.

Why do we need enzymes?

At 37°C, without enzymes chemical reactions in your body would happen too slowly to keep you alive.

A higher body temperature could speed up the chemical reactions in your body. But higher temperatures damage human cells. For a higher body temperature you would need a lot more food to fuel respiration.

However, enzymes can speed up the rate of a reaction by up to 10 000 000 000 times. We need enzymes to live.

Shrews are very small. They lose heat quickly. To release enough energy to keep a steady body temperature, they have to eat 75% of their body mass in food each day. If not, they die within 2–3 hours.

Around 80% of the energy from your food is used for keeping warm. If you had to maintain a higher body temperature, you would have to spend a lot more time eating.

Questions

1. Write down:
   a. what enzymes are made of
   b. what enzymes do
2. Explain how an enzyme works. Use the words 'active site' and 'lock-and-key model' in your answer.
3. The enzyme amylase breaks down **starch** to sugar; the enzyme catalase breaks down hydrogen peroxide to water and oxygen. Explain why catalase does not break down starch.

Summary box
- **Enzymes are catalysts – they speed up chemical reactions.**
- **The lock-and-key model explains how enzymes work.**
- **Only the right shaped molecule will fit into the enzymes active site.**

C — Keeping the best conditions for enzymes

Find out about
- how temperature and pH affect enzymes

How does temperature affect enzymes?

At low temperatures, enzyme reactions get faster if the temperature is increased. But above a certain temperature the reaction stops. This is because enzymes are proteins. Higher temperatures change an enzyme's shape so that it no longer works. The diagram on the left explains what happens using the lock-and-key model.

Enzymes stop working at high temperatures

High temperatures change the shape of an enzyme. They do not destroy it completely. But even when an enzyme cools down, it does not go back to its original shape. Like cooked egg white, the protein cannot be changed back.

Why does our body keep a constant temperature of 37 °C?

The temperature at which an enzyme works best is called its optimum temperature. It is a temperature too low to denature the enzyme, but not so low that the reaction slows down. Enzymes in humans work best at around 37 °C. Some organisms have cells and enzymes adapted to different temperatures.

Bacteria living in hot springs have enzymes that withstand high temperatures.

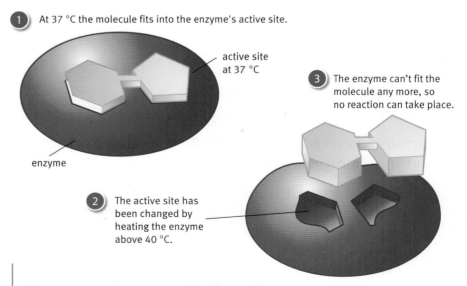

1. At 37 °C the molecule fits into the enzyme's active site.
2. The active site has been changed by heating the enzyme above 40 °C.
3. The enzyme can't fit the molecule any more, so no reaction can take place.

How an enzyme reaction can be stopped by a rise in temperature.

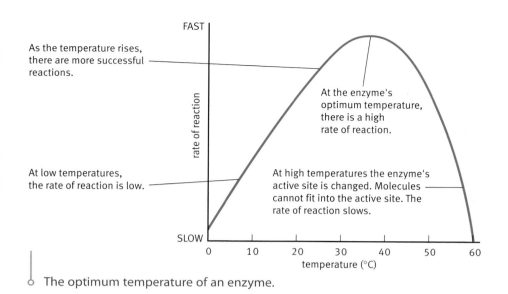

The optimum temperature of an enzyme.

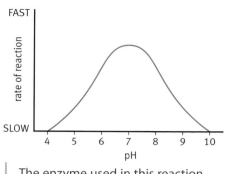

The enzyme used in this reaction has an optimum pH of 7.0.

pH also affects enzymes

The shape of an enzyme's active site is also affected by pH. Every enzyme has an optimum pH at which it works best.

Enzyme	What it does	Optimum pH
salivary amylase	breaks down starch to sugar	4.8
pepsin	breaks down proteins into short chains of amino acids	2.0
catalase	breaks down hydrogen peroxide into water and oxygen	7.6

Questions

1 Describe how an enzyme-catalysed reaction changes:
   a as the temperature increases at low temperatures
   b at higher temperatures.
2 What is meant by an enzyme's optimum temperature?
3 Enzymes from microorganisms, such as bacteria and fungi, make food decay. Why does food stay fresh for months in a freezer?

Summary box
- Increasing the temperature speeds up a reaction.
- Enzymes are specially shaped proteins. Their shape is damaged by high temperatures and pH changes.
- If an enzyme's active site is damaged, molecules will no longer fit into it. The reaction will slow down or stop.
- Each enzyme works best at its optimum temperature and pH.

D | Enzymes at work in plants

Find out about

- how photosynthesis captures energy for life on Earth
- what happens to the glucose made by photosynthesis

Enzymes speed up reactions in animals and plants too. We rely on plants because they can make their own food from simple things using the Sun's energy. Photosynthesis makes food molecules and energy available to living things through food chains.

Plants make glucose and oxygen by photosynthesis

Plants, and some microorganisms such as phytoplankton, can make glucose (sugar) molecules from simple carbon dioxide gas and water. They make oxygen too as a waste product. Sunlight energy drives the reaction. Plants use a green pigment called **chlorophyll** to capture the Sun's energy. Chlorophyll is found inside tiny **chloroplasts** inside the leaf cells.

What happens during photosynthesis?

The chemical equation for photosynthesis is:

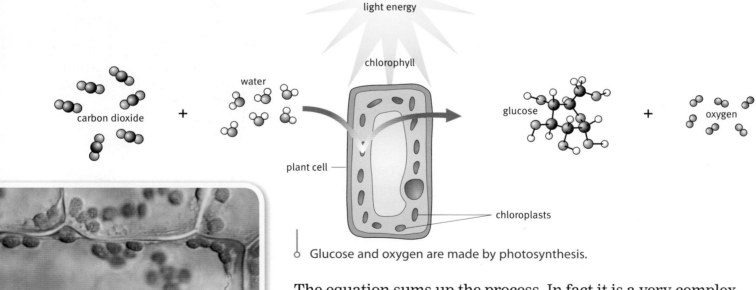

Glucose and oxygen are made by photosynthesis.

Chloroplasts contain the green pigment chlorophyll and the enzymes that are needed for photosynthesis ($\times 2000$).

The equation sums up the process. In fact it is a very complex process that happens in lots of smaller steps. Each step is helped by special enzymes.

A glucose molecule is made up of carbon, hydrogen, and oxygen atoms. So glucose is called a carbohydrate. On a warm day a large tree can make 2000 kg of glucose if it has enough water, fresh air, and sunlight.

Using glucose from photosynthesis

Glucose made during photosynthesis is used by plant cells in three ways.

(1) Molecules for growth

Some plants grow very fast. Glucose can be converted into different molecules needed for cell growth. For example, the plant's cell wall is made of a strong material called **cellulose**. Proteins are needed to make new cells. Proteins and cellulose are polymers. Polymers are large molecules made up of thousands of smaller molecules linked together. Cellulose is a polymer of glucose, for example.

(2) Storing energy in starch molecules

Sometimes photosynthesis produces glucose faster than the plant needs it. This extra glucose is changed into starch. Starch is a storage polymer – it is made of thousands of glucose molecules joined together. When extra glucose is needed, the starch can be changed back into glucose.

(3) Releasing energy in respiration

Plant cells use glucose in respiration. The glucose molecules are broken down, releasing the energy stored in the molecules.

This energy is used to power chemical reactions in the cells, such as changing glucose to cellulose, starch, or proteins.

Questions

1. Write down the word equation that sums up photosynthesis.
2. Draw a diagram to show the flow of chemicals in and out of leaves during photosynthesis.
3. Glucose from photosynthesis has three roles in the plant cell. Describe what these are.

Summary box

- Plants make glucose and oxygen by photosynthesis.
- The simple ingredients are water and carbon dioxide gas.
- Energy from the Sun is captured by chlorophyll to drive the reaction.
- We rely on plants for food and oxygen.

A close-up of a plant's cell wall showing strong cellulose fibres. Cellulose molecules are polymers – long straight chains of glucose molecules.

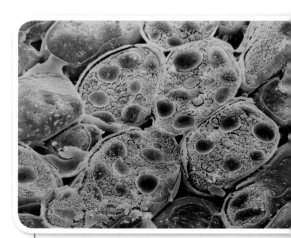

Starch grains inside these plant cells store glucose as starch ($\times$ 200).

E — Diffusion and gas exchange in plants

Find out about

- how chemicals move in and out of cells
- how carbon dioxide and oxygen move in and out of a leaf during photosynthesis

Molecules move in and out of cells all the time. Cells need a constant supply of raw materials for chemical reactions, and waste needs to be removed.

Photosynthesis in a plant leaf cell uses carbon dioxide and produces waste oxygen. Movement of these molecules takes place by the process of **diffusion**.

Molecules move by diffusion

Molecules in gases and liquids move about randomly. They collide with each other and change direction. This makes them spread out.

① Water just poured onto tea bag

To start with, dissolved molecules from the tea are concentrated close to the tea bag. There are no tea molecules in the rest of the hot water.

② About 30 seconds later

The tea molecules move from where they are highly concentrated into regions where they are less concentrated.

③ About 2 minutes later

After a few minutes the tea molecules will spread evenly throughout the cup.

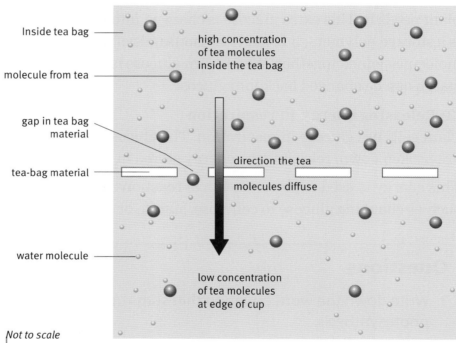

Not to scale

Dissolved tea molecules diffuse out of the tea bag and into the surrounding water.

Molecules that are free to move won't stay bunched together for long. The tea molecules spread apart from each other. Finally, all the water is the same brown colour. We say the molecules diffuse from areas of their high concentration to areas of low concentration.

The tea molecules will spread out uniformly, even if the cup of tea is not stirred. Diffusion is a passive process – no extra energy is needed.

This swimmer's cells need oxygen and glucose for respiration. They must get rid of carbon dioxide. These molecules move in and out of cells by diffusion.

Photosynthesis, diffusion, and gas exchange in leaves

Plants make their own food by photosynthesis in the leaves. Leaves are well designed to let gases diffuse in and out.

Plant cells are designed to carry out photosynthesis. They have structures that play a role in the chemical reactions of photosynthesis.

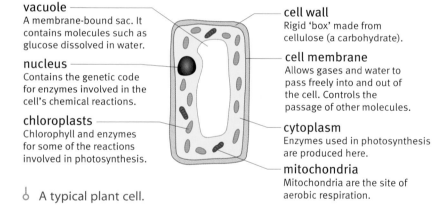

vacuole
A membrane-bound sac. It contains molecules such as glucose dissolved in water.

nucleus
Contains the genetic code for enzymes involved in the cell's chemical reactions.

chloroplasts
Chlorophyll and enzymes for some of the reactions involved in photosynthesis.

cell wall
Rigid 'box' made from cellulose (a carbohydrate).

cell membrane
Allows gases and water to pass freely into and out of the cell. Controls the passage of other molecules.

cytoplasm
Enzymes used in photosynthesis are produced here.

mitochondria
Mitochondria are the site of aerobic respiration.

A typical plant cell.

The underside of a leaf has thousands of tiny pores. These allow carbon dioxide into the leaf and oxygen out. Diffusion drives these gases in opposite directions. Each gas moves from high to low concentrations.

Questions

1 Write down a definition for diffusion.
2 Name three chemicals that move in and out of cells by diffusion.

Summary box

- **Molecules in liquids and gases:**
  - move about all the time
  - spread apart from each other
  - move from areas of high concentration to areas of low concentration.
- **Diffusion is the movement of molecules from an area of their higher concentration to an area of their lower concentration.**
- **Leaves have tiny pores to let gases diffuse in and out.**

F | Osmosis

Find out about

- the movement of water molecules by osmosis
- why cells need a steady water balance
- why plant cells store glucose as starch
- balancephotosynthesis

Summary box

- Osmosis is the flow of water from a dilute to a more concentrated solution across a partially permeable membrane.

Osmosis is a specific type of diffusion. It is the process that moves water molecules into and out of cells. The cell membrane is important in osmosis.

Cell membranes are partially permeable

Cell membranes let some molecules through but block others. Tiny channels in the membrane allow small molecules, like water, to travel through them. Larger molecules are too big and cannot get through. Cell membranes are **partially permeable membranes**.

Diagrams **a** and **b** below show how water molecules move during osmosis. In this example the membrane allows water through but the glucose molecules are too big to get through.

More water molecules move away from an area of higher concentration of free water molecules. Think of it as diffusion of water. This overall flow of water from a dilute to a more concentrated solution across a partially permeable membrane is called osmosis.

Key

 partially permeable membrane allows some molecules through and acts as a barrier to others

 glucose molecule

○ water molecule

 water molecules associated with glucose molecule (these molecules are not free to move by osmosis)

(*Note:* In these diagrams, the circles represent molecules, not individual atoms. Cell membranes are also made of molecules.)

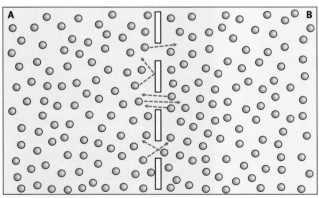

a This membrane is separating water molecules. Water molecules move at random. As many pass from left (A) to right (B) as pass from right (B) to left (A).

Not to scale

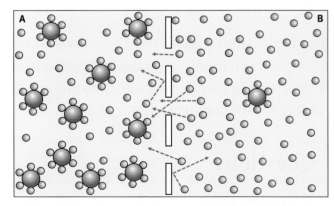

b This membrane is separating two glucose solutions. There are more free water molecules and fewer glucose molecules on the right (B). In other words, water molecules that are free to move by osmosis are in higher concentration on the right. So there is overall movement of water from right (B) to left (A).

Osmosis in plant cells

Plants do not have skeletons to give them support. They must keep their shape by having firm cells that are 'pumped up' with water. Water molecules move from the soil to the plant roots and then to the leaves by osmosis. Osmosis drives the uptake of water by plant roots and determines how water passes from one cell to another throughout the whole plant.

If plant cells take in too much water they bulge and become stretched. Their strong cell wall prevents them from bursting. If plant cells lose too much water they shrink. Plant cells need to keep just the right amount of water inside. Look at the pictures to see what happens to a plant if it does not have enough water.

The photographs show the same plant. The plant on the right is shown after it has not been watered for 10 days. Water is important for keeping the leaf cells firm and 'pumped up'.

Glucose is stored as starch

Plants make glucose during photosynthesis. The glucose is transported from the leaves to other cells where it is stored until it is needed for respiration. This poses a problem for the plant cells that store the glucose. They will take in too much water. The water would move from a dilute solution surrounding the cells into the more concentrated glucose solution in the cells. To overcome this problem, glucose is stored as starch.

Large carbohydrates like starch are **insoluble**. They have very little effect on the concentration of the solutions in a plant cell.

Summary box
- Water enters and leaves cells by osmosis.
- Living things need to keep a steady water balance to be healthy.
- Plants store glucose as insoluble starch.

Questions

1. Explain how diffusion lets you smell the vinegar from fish and chips on a plate in front of you.
2. Explain how sugars, made in photosynthesis, get into the phloem vessels that transport them to the rest of the plant.
3. Explain what is meant by a partially permeable membrane.
4. Write down a definition of osmosis.
5. A student put a raisin (a dried grape) into a glass of water. They noticed that the raisin expanded and swelled. Explain this observation. Include a diagram to show the movement of water.
6. Explain why glucose is stored in the form of starch in plant cells.

G — Minerals from the soil

Find out about

- why plants need minerals
- how minerals are absorbed in the roots

Nitrates contain this group of atoms:

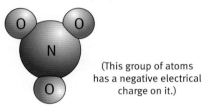

(This group of atoms has a negative electrical charge on it.)

Nitrate ions are found dissolved in soil water, and in rivers and seas.

Plants capture energy from sunlight during photosynthesis. This energy builds glucose for respiration. Glucose also supplies the raw materials needed to make other molecules like proteins, fats, and DNA. These molecules need elements contained in minerals from the soil.

Making proteins needs nitrogen

Proteins are long chains of amino acids. To make amino acids, carbon, hydrogen, and oxygen atoms from glucose must be combined with nitrogen.

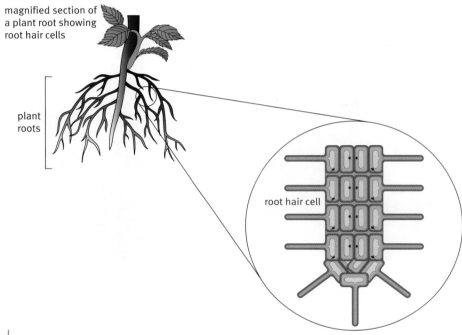

Plant roots have a very large surface area to help them absorb water and minerals. Tiny root hair cells take in minerals, including nitrates.

Most of the Earth's nitrogen is in the air, but plants mainly take in nitrogen from the soil as **nitrates**. Nitrates enter the plant when water is absorbed into the root cells by osmosis. A range of minerals are absorbed by **root hair cells**.

Mineral in fertiliser	Used in plant to make
Nitrate	proteins
Magnesium	chlorophyll
Phosphate	DNA

The rate of photosynthesis

The conditions inside the greenhouse in the photograph below are kept under very careful control by the farmer. The tomato plants growing here have the best conditions for photosynthesis. The yield from the tomato plants will be as high as possible. Yield is the amount of product the farmer has to sell.

Faster photosynthesis

All reactions speed up when the temperature rises. The greenhouse is kept warm at 26 °C. This is the best temperature for photosynthesis to take place in these plants.

Other things will speed up the rate of photosynthesis. Light energy drives photosynthesis. Increasing the amount of light a plant receives increases the rate of photosynthesis.

> **Find out about**
> ✓ what limits the rate of photosynthesis

Intensive tomato farming takes place all year round in this greenhouse.

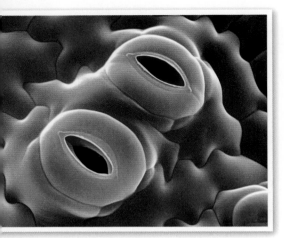

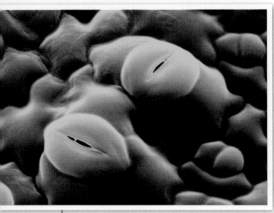

Top: These pores called stomata on the underside of leaves open to allow gases to move in and out of the leaf. *Bottom:* They close to conserve water (× 400).

The diagram below shows an experiment to investigate how changing **light intensity** affects the **rate of photosynthesis** in a piece of pondweed. The results from the experiment are shown in the graph:

- at low light levels, increasing the amount of light increases the rate of photosynthesis
- at a certain point increasing the amount of light has no more effect on the rate of photosynthesis.

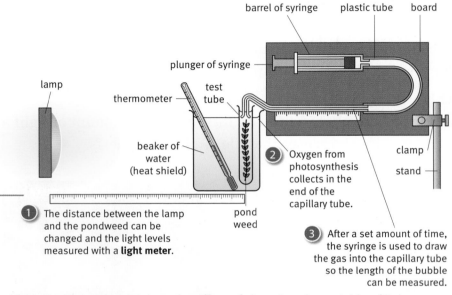

① The distance between the lamp and the pondweed can be changed and the light levels measured with a **light meter**.

② Oxygen from photosynthesis collects in the end of the capillary tube.

③ After a set amount of time, the syringe is used to draw the gas into the capillary tube so the length of the bubble can be measured.

This experiment investigates the effect of changing the variable of light intensity on the output variable rate of photosynthesis.

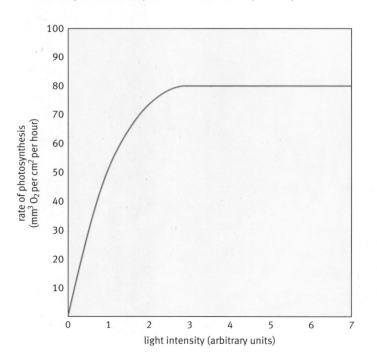

Why does the rate not keep on rising?

Photosynthesis needs more than light energy. Extra light makes no difference to the rate of photosynthesis if something else is in short supply. The plant must have enough carbon dioxide, water, and chlorophyll to use the energy to the full. The temperature must also be high enough. Increasing the light intensity stops having an effect on the rate of photosynthesis because one of these other factors is in short supply. This factor is called the **limiting factor**.

Limiting factors

The graph below shows the effect of increasing light intensity on the rate of photosynthesis. Two different carbon dioxide concentrations have been plotted. At 0.04% CO_2, more light increases the rate of photosynthesis up to a point, until light is no longer the limiting factor. Increasing the CO_2 level to 0.4% makes the rate of photosynthesis higher – carbon dioxide must have been the limiting factor. But even this graph levels off as another factor becomes in short supply.

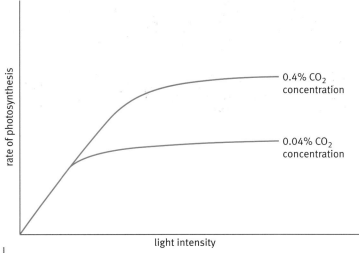

At the higher carbon dioxide concentration of 0.4%, photosynthesis takes place faster. But the rate still levels off. Another factor must be limiting photosynthesis.

Carbon dioxide levels in the greenhouse

Carbon dioxide forms 0.04% of normal air. The levels in the tomato greenhouse are raised to 0.1% to speed up photosynthesis. Raising the concentration higher than this is not cost effective.

Summary box
- Reactions speed up:
  - when you increase the temperature
  - when you increase the concentration of reactants.
- Photosynthesis is faster when there is the right amount of warmth, light, and carbon dioxide.
- A limiting factor is something that is in short supply that holds back the rate of photosynthesis.

Questions

1. Write down three factors that can affect the rate of photosynthesis.
2. Explain what is meant by a limiting factor.
3. Suggest a factor that could be limiting bluebell flowers growing on a woodland floor in spring.

Environments and adaptations

Find out about

- how environmental conditions affect the plants that are able to grow
- ways to survey plants in a location

Summary box

- To grow well all plants need the right amount of light, warmth, water, and minerals.
- These factors are different in different habitats.
- Different plants grow best in different conditions.
- Percentage plant cover can be surveyed using quadrats placed at random.
- A transect is a survey along a straight line.

Why do different plants grow in different locations?

Don't plants just grow where we plant them? Well that is the case in gardens and parks, but not in natural ecosystems. The great variety of plants has evolved to take advantage of different **habitats** all over the Earth.

All plants need minerals, water, and light in just the right amount to be able to grow well.

Some plants grow well in shade and others need bright light. Some plants need plenty of water and others can survive in deserts.

The ground in this woodland is quite shady. Only certain plants will be able to grow here.

Investigating different habitats

To understand why plants grow in particular places, factors like soil pH, temperature, light levels, and the availability of water are measured. Enough individual **samples** must be taken to get a true picture of what the conditions are like.

A square grid, called a **quadrat**, is used to survey the plants in a square metre. The quadrat is placed on the ground. Plants within the quadrat are identified and counted. An identification key can help to name the organisms. The key has descriptions or pictures that can be compared to the specimen. Plant growth is often recorded as percentage (%) cover.

The placing of the quadrat in the area being investigated is **random**. Placing quadrats randomly removes bias. This allows reliable comparisons between different locations. Recording the plants in a quadrat involves accurate identification of each species.

Sometimes samples are taken at regular intervals along a straight line called a **transect**. This is useful when looking at how the types of plants change gradually from one area to another, for example, when moving from the shaded part of a wood into an open field.

A light meter could be used to accurately measure light intensity. The sensor should be held at the same angle and the readings taken one after the other so as to compare like with like.

This scientist is using a quadrat to record the plants growing.

Comparing where different plants grow

The results of a survey of bluebell plants are presented in the table below. Two areas were compared – one in shady woodland and the other in the middle of a field.

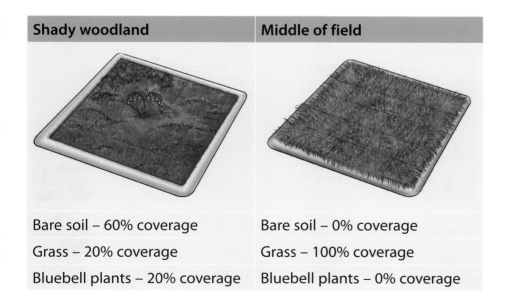

Shady woodland	Middle of field
Bare soil – 60% coverage	Bare soil – 0% coverage
Grass – 20% coverage	Grass – 100% coverage
Bluebell plants – 20% coverage	Bluebell plants – 0% coverage

The results show that bluebell plants can survive in the shade. In the field, the bluebells cannot compete with the grass.

Questions

1. What do plants need to be able to grow?
2. What are the different environmental conditions that could affect plant growth?
3. Suggest how you could make sure that plant and animal samples taken from a location give a reliable indication of the conditions in the area.

J Energy for life

Find out about
- aerobic respiration in plant and animal cells

Photosynthesis makes glucose from carbon dioxide and water. This captures the energy held in sunlight and converts it into energy held in glucose. Cells release the energy in glucose in a process called respiration. This is a complex chain of chemical reactions that happen in the cells of all living things.

Cells need energy for movement, growth, and repair.

Aerobic respiration uses oxygen

The cells in your body need a constant supply of energy. The food you eat provides you with molecules to make new cells. But it is also a store of chemical energy. This is converted by respiration into energy your cells can use.

Most of your energy comes from **aerobic respiration**. Aerobic respiration takes place in animal and plant cells, and some microorganisms. During aerobic respiration, glucose from your food reacts with oxygen from the air that you have breathed. The reactions release energy from the glucose. Respiration can be summarised by the following equation:

$$\text{glucose} + \text{oxygen} \rightarrow \text{carbon dioxide} + \text{water} (+\text{energy released})$$

Respiration is a complex chain of reactions. This equation sums it up.

This girl is using energy for many different actions – such as moving, growing, growing and repairing her tissues.

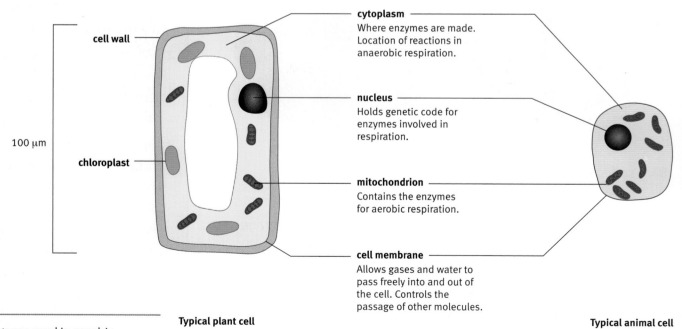

Structures used in aerobic respiration in plants and animal cells.

Typical plant cell (100 μm)
- **cell wall**
- **chloroplast**
- **cytoplasm** — Where enzymes are made. Location of reactions in anaerobic respiration.
- **nucleus** — Holds genetic code for enzymes involved in respiration.
- **mitochondrion** — Contains the enzymes for aerobic respiration.
- **cell membrane** — Allows gases and water to pass freely into and out of the cell. Controls the passage of other molecules.

Typical animal cell

B4: THE PROCESSES OF LIFE

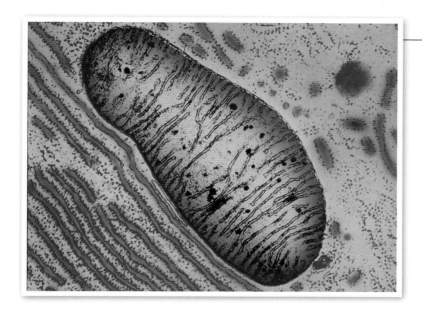

Some of the reactions for respiration take place in the cell cytoplasm, and many happen inside **mitochondria**. This electron micrograph shows a single mitochondrion inside a cell ($\times 64\,000$).

What happens to the energy from respiration?

Respiration releases energy from glucose. This energy will be needed by many processes in the cell.

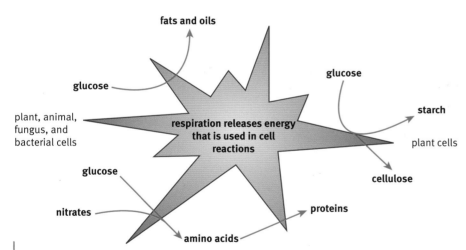

The energy from respiration can be used to join up small molecules to make large ones. These include **polymers** of glucose (starch and cellulose) and proteins, fats, and oils.

Summary box
- Cells need energy for lots of processes, including movement, growth, and repair.
- Respiration releases the chemical energy stored in glucose.
- Aerobic respiration uses oxygen from the air and releases energy, carbon dioxide gas, and water.
- Energy from respiration is held in ATP in the short term for easy use.
- Aerobic respiration takes place inside tiny mitochondria inside each cell.

Questions

1. Write down three processes your body needs energy for.

2. a Write down the word equation for aerobic respiration.
   b Label the equation to show where the reactants come from, and what happens to the products.

J: ENERGY FOR LIFE

K Energy without oxygen

Find out about

- anaerobic respiration in humans and other organisms

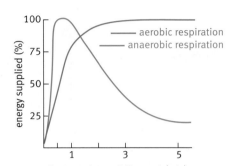

Anaerobic respiration provides a burst of energy for quick sprints. Races over 400 m need a more steady pace with energy coming from aerobic respiration.

Animals use a different type of respiration for short bursts of intense energy, for example, when a predator runs after its prey and the prey runs for its life. In these situations the animals' muscles cannot get oxygen quickly enough for aerobic respiration. So they switch to a form of respiration that does not need oxygen. It is called **anaerobic respiration**.

glucose $\longrightarrow$ lactic acid (+ energy released)

Anaerobic respiration in animals and some microbial cells is summarised by this equation.

In animals, anaerobic respiration can only be used for a short period of time. It releases much less energy from each gram of glucose than aerobic respiration. Also, the waste product **lactic acid** is toxic in large amounts. If it builds up in muscles, it makes them feel tired and sore.

Plants can also use anaerobic respiration, for example, when the roots are in waterlogged soils, or the seeds germinate underground.

The 100 m sprint takes about 10–11 seconds. The heart and lungs cannot increase oxygen supply to the muscles fast enough, so most of the energy required during the short race comes from anaerobic respiration.

Germinating seeds respire anaerobically.

Anaerobic respiration in bacteria

Some bacteria can use anaerobic respiration. They can cause stinking infections in deep puncture wounds. However, we can put anaerobic bacteria to work. The lactic acid that they make gives yoghurt and cheese its tangy taste.

Anaerobic respiration in yeast

Yeast is a single-celled fungus. Yeast has been used to make wine, beer, and bread for thousands of years.

glucose → ethanol + carbon dioxide (+ energy released)

Anaerobic respiration in plants and some microorganisms including yeast is summarised by this equation.

Yeast turns the sugars in grape juice into ethanol (alcohol). Sparkling wine is fizzy because carbon dioxide gas is trapped in the bottle.

Bread 'rises' because carbon dioxide gas bubbles expand inside the dough. The alcohol is driven out when bread is baked in the oven. This gives baking bread a nice smell.

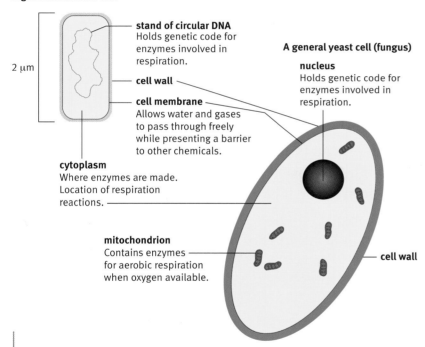

Structure of a typical bacteria and yeast cell.

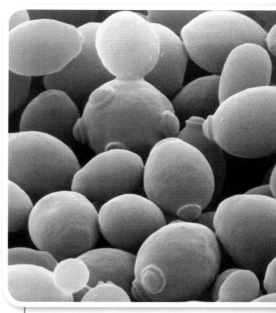

Yeast cells can respire anaerobically until ethanol (alcohol) builds up and becomes too toxic.

Summary box
- Anaerobic respiration releases energy without oxygen.
- Anaerobic respiration releases less energy than aerobic respiration.
- Anaerobic respiration is used:
  - in fast sprints
  - in waterlogged plant roots
  - in bacteria to make lactic acid
  - in yeast to make ethanol and carbon dioxide.

Questions

1. Describe conditions where anaerobic respiration is useful to:
   a human beings   b plants
2. Give an example of anaerobic respiration that people use to make a useful product.

L — Useful products from respiration

Find out about
- how bioethanol is made from sugar
- how farm manure can be used to get a valuable fuel

Microorganisms can help us make food and drinks like cheese, yoghurt, bread, and wine. They can also make valuable fuels on a large scale.

Ethanol from yeast can fuel cars

Yeast is a single-celled fungus. During anaerobic (oxygen-free) respiration, it makes ethanol (alcohol) and carbon dioxide. This is how we brew beer and make wine.

Ethanol contains a large amount of stored chemical energy. It can power car engines. Large amounts of fuel ethanol – called **bioethanol** – can be made from sugar cane, sugar beet, maize, and wheat. Yeast respires anaerobically in huge tanks. This process is called **fermentation**. The alcohol is separated from the sugary water by distillation.

Biofuels and sustainability

Crude oil is running out. Bioethanol is made from renewable plant sources. Farmers can plant more sugar crops if they want to make more fuel. Bioethanol is called a **sustainable** fuel.

However, biofuels can be controversial. Land may be used to grow crops for fuel instead of food for people. In some areas, forest is cut down so the land can be used to grow crops for fuel. Producers are now developing ways of using non-food crops and algae to end the need to use food crops and valuable land.

Carbon dioxide, a by-product of the anaerobic respiration of yeast, has caused these loaves of bread to rise.

This factory produces bioethanol (alcohol) to fuel cars.

Biogas from waste

Biogas is a fuel obtained from animal manure or sewage. The bacteria respire anaerobically in airtight tanks. Bacteria break down the organic material in the manure and produce methane gas and some carbon dioxide. This can be used as a fuel to heat buildings and run electricity generators.

Biogas is made on a small scale in some developing countries. It brings many benefits:
- Collecting manure and sewage encourages hygiene.
- Everyone shares the biogas, which saves firewood.
- Biogas cookers are less smoky than open fires, which is healthier.
- The digested manure can fertilise crops.

Bacteria in biogas fermenters produce methane gas and carbon dioxide during anaerobic respiration.

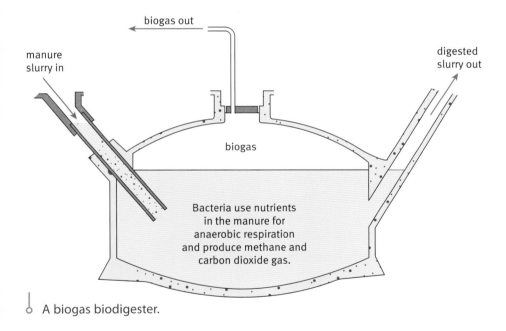

A biogas biodigester.

These airtight tanks at a sewage works make biogas.

Summary box
- Anaerobic respiration can make fuels on a large scale.
- Bioethanol (alcohol) is made by yeast from sugary plant material.
- Biogas (methane) is made by bacteria from sewage.
- Biofuels are sustainable – they won't run out.
- Biofuels from algae and waste will save food crops and valuable land.

Questions

1. Why is bioethanol said to be a renewable resource?
2. Describe the process of anaerobic respiration in yeast. Compare it to anaerobic respiration in animal cells.
3. Draw a flow diagram to show how energy is transferred from sunlight into the methane in biogas. Include details of the processes happening at each stage.

Science Explanations

All living things are made up of cells. Biological processes such as photosynthesis and respiration take place in cells. These processes involve chemical reactions that are speeded up by enzymes.

You should know:

- that some chemical reactions in cells require energy; these reactions include muscle contraction and the synthesis of large molecules
- that respiration is a series of chemical reactions in plant, animal, and microbial cells, which release energy by breaking down food molecules
- what enzymes are and how they speed up chemical reactions in living organisms
- that cells make enzymes by following the instructions carried in our genes
- why the model describing how enzymes recognise molecules is called the lock-and-key model
- the conditions needed by enzymes to work at their optimum rate
- that photosynthesis uses energy from sunlight to make glucose, by joining carbon dioxide and water and releasing oxygen as a waste product
- that chlorophyll absorbs light energy
- that glucose may be converted into other chemicals needed for plant growth
- how to use equipment such as light meters, quadrats, and identification keys to investigate the effect of light on plants
- how to take a transect in fieldwork
- that minerals are taken up by plant roots, and what they are used for in plants
- that diffusion is the passive overall movement of molecules from a region of higher concentration to a region of lower concentration
- how osmosis is the overall movement of water by diffusion from a dilute to a more concentrated solution through a partially permeable membrane
- that nitrates are absorbed by plant roots to make some chemicals needed by cells
- that aerobic respiration breaks down glucose in the presence of oxygen, releasing energy, carbon dioxide, and water; it takes place in animal and plant cells and some microorganisms
- that anaerobic respiration takes place in animal, plant, and some microbial cells
- how energy is released from glucose without oxygen during anaerobic respiration – animal cells form lactic acid; plant cells and yeast produce carbon dioxide and ethanol
- the structures and functions of typical plant, animal, and microbial cells
- how we use the anaerobic respiration of microorganisms to make biogas, bread, and alcohol.

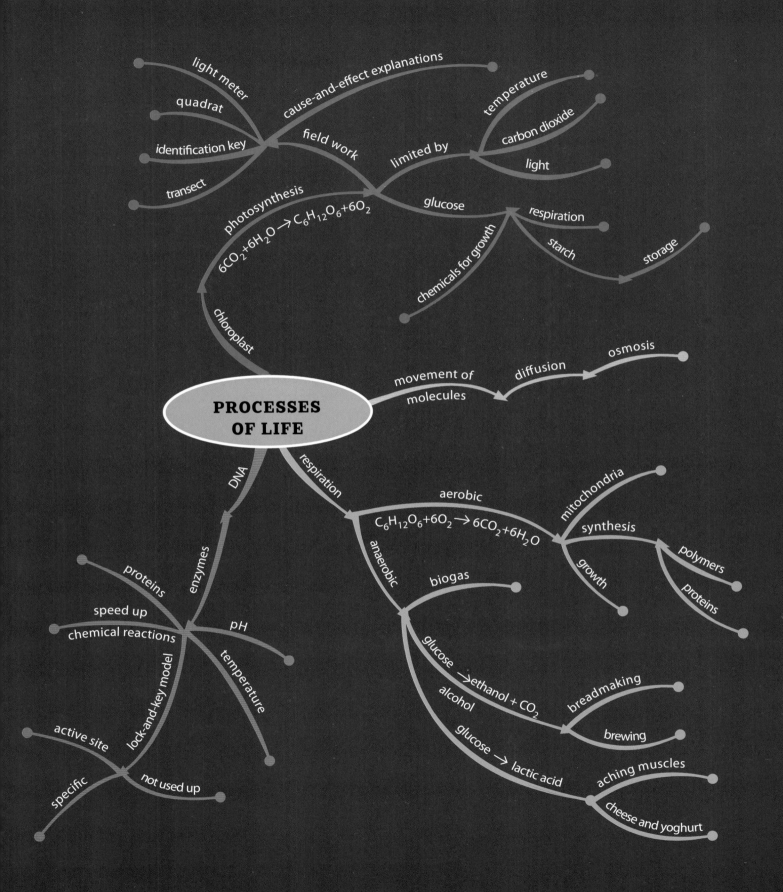

Ideas about Science

This module helps you develop an understanding of life processes. You will also learn more about how scientists explain cause and effect, and how they investigate relationships between factors.

If you take several measurements of the same quantity, these are likely to vary. This may be because:
- you are measuring several individual examples, such as oxygen produced from different samples of pond weed
- the quantity you are measuring is varying, for example, the number of bluebells growing in different areas of a wood
- the limitations of the measuring equipment or because of the way you use the equipment.

The best estimate of a true value of a quantity is the mean of several measurements. The true value lies in the spread of values in a set of repeat measurements.
- A measurement may be an outlier if it lies outside the range of the other values in a set of repeat measurements.
- When comparing information on plants growing in different places, a difference between their means is likely to be real if their ranges do not overlap.
- A correlation shows a link between a factor and an outcome, for example, as light intensity increases, the rate of photosynthesis increases.
- A correlation does not always mean that the factor being changed causes the outcome.

Scientists often think about processes in terms of factors that may affect an outcome (or outcome variable). When you investigate the relationship between a factor and an outcome, you need to keep the same (control) all the other factors that you think might affect the outcome of the investigation. This is called a 'fair test'.

You should be able to:
- identify input variables for photosynthesis
- recall that an enzyme works best at its optimum temperature and pH

You also need to outline why these variables must be controlled in an investigation, and to consider how you would perform a fair test.

When you plan an investigation you need to identify the likely effect of a factor on an outcome. You also need to know that if you don't control the variables in an investigation, the investigation will be flawed.

You should understand why plants grow in particular places. You could measure factors like soil pH, temperature, light intensity, and water. Compare this data with what you know about the plants growing in an area, and what plants need. Think about cause-and-effect explanations for what you have found out.
- Enough individual samples must be taken to get a true picture of the conditions.

You should look for ways of collecting reproducible data and be able to outline why repeating measurements leads to a better estimate of the quantity measured. Here are two examples.
- A light meter can be used to accurately measure light intensity. In a well-planned experiment the sensor should be held at the same angle for each reading. Time of day and shading from trees and bushes should be taken into account.
- One way of sampling using quadrats is to position quadrats randomly in the area being measured. This removes bias. Recording the quadrat data involves accurate identification of each species.

Review Questions

1
a What are enzymes made from?
b What do enzymes do?
c Enzymes and some molecules that are the correct shape can fit together.
Which model explains this?
Choose the correct answer.

    enzyme-and-molecule model
    lock-and-key model
    puzzle-shaped model

d What would happen to the speed of an enzyme reaction if the temperature was raised very high?

2 Alex draws a set of diagrams to show what happens during a reaction involving an enzyme.

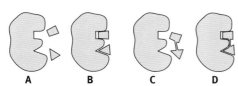

The stages are not drawn in the correct order.
Put the stages in the correct order.
The last one has been done for you.

3 Photosynthesis takes place in green plants.
a What would happen to the rate (speed) of photosynthesis if:
  i the amount of light was decreased?
  ii the temperature was to get lower?
  iii extra carbon dioxide gas was added?

b This equipment shown below can be used to investigate the rate of photosynthesis.
  i Describe how you would use the equipment to investigate how the light intensity (brightness) affects the rate of photosynthesis.
  ii Describe two things that you could do to make sure that this was a fair test.

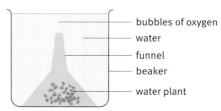

4 Steve is an athlete. He runs marathons. During most of the race he releases energy aerobically in his muscles.
a Describe what happens in aerobic respiration.
b During the sprint to the finish his muscles use anaerobic respiration. Describe what happens in anaerobic respiration.

5 Write a definition of:
i osmosis
ii diffusion.

Give an example of where each might happen.

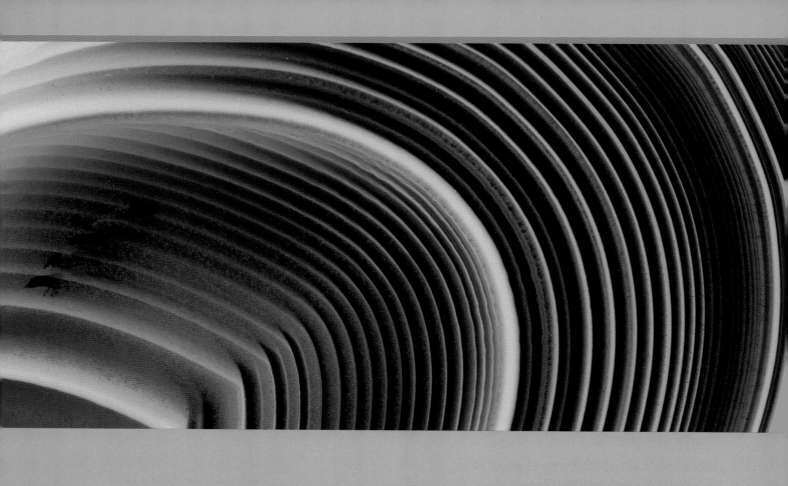

C4 chemical patterns

C4: CHEMICAL PATTERNS

Why study chemical patterns?

The periodic table is important in chemistry because it helps to make sense of all the information about the elements and their compounds. The table offers a framework that can give meaning to all the facts about properties and reactions.

What you already know

- Elements are made up of just one type of atom.
- A molecule is a group of atoms chemically joined together.
- During a chemical reaction atoms are conserved.
- The properties of the reactants and products of chemical changes are different.
- There are patterns in the way chemicals react.
- Acids neutralise alkalis to form salts.
- Chlorine can be used to kill microorganisms.
- A chemical change caused by the flow of an electric current is called electrolysis.

Find out about

- the chemistry of some very reactive elements
- the patterns in the periodic table
- how scientists learn about the insides of atoms
- how atoms become charged and turn into ions.

The Science

Chemists use ideas about the structure of atoms to explain the periodic table and the properties of the elements. Spectroscopy led to the discovery of new elements. Today, spectroscopy is used to study chemicals and chemical reactions.

Ideas about Science

For a scientific explanation to be accepted and become a theory, it must be tested by other scientists to check it is supported by evidence and can be used to make predictions. Atomic theory helps to explain patterns in the periodic table.

A | The periodic table

Find out about

- the development of the periodic table
- groups and periods

Germanium – one of Mendeleev's missing elements. He predicted what its properties would be.

Questions

1. What did Döbereiner notice about the properties of elements?
2. What did Newlands do with the elements and what did he notice?
3. When Mendeleev lined up the elements in order he left gaps. What did these gaps represent?
4. What new findings encouraged other scientists to accept Mendeleev's ideas?

Looking for patterns

At the beginning of the 1800s, only about 30 elements were known, but by the end of that century almost all of the stable elements found on Earth had been discovered.

There were lots of elements, with a wide range of properties. Chemists began to look for patterns in their properties to try to put them into a useful order.

Relative atomic masses

In those days scientists could not measure the actual masses of atoms. Instead they compared the masses of atoms with the mass of the lightest one, hydrogen. This is called the **relative atomic mass**.

In the early 1800s, Johann Döbereiner, a German scientist, noticed that there were several groups of three elements with similar properties. Almost 50 years later an English chemist, John Newlands, arranged the elements that were known at the time in order of their relative atomic masses. He saw that every eighth element had similar properties. This seemed only to work for the first 16 of the known elements. Other scientists did not accept these ideas, because they didn't include all the known elements.

Mendeleev

Dmitri Mendeleev, a Russian scientist, showed that it is possible to come up with patterns using all the elements when they are lined up in order of their relative atomic mass. Mendeleev's inspiration was to realise that not all of the elements had yet been discovered. He left gaps for missing elements when necessary to produce a sensible pattern.

Mendeleev predicted the properties of the missing elements. When a new element was discovered that fitted his predictions, his ideas were accepted by other scientists.

The periodic table now

In the modern periodic table, the elements are arranged in rows, one above the other. Each row is a **period**.

There are repeating patterns in the periodic table. The most obvious is from metals on the left to non-metals on the right. Every period starts with a very reactive metal in Group 1 and ends with an unreactive gas in Group 0.

Elements with similar properties fall into a column. Each column is a **group** of similar elements.

Summary box
- Döbereiner and Newlands noticed patterns in elements.
- Mendeleev predicted properties of new elements using his periodic table.
- A group is a column in the periodic table; a period is a row.

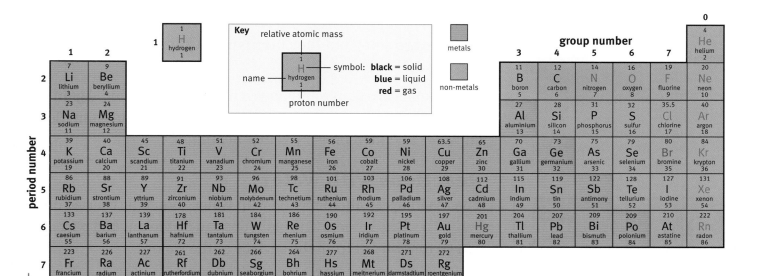

The periodic table. Over three-quarters of the elements are metals. They lie to the left of the table. (Elements with proton numbers 58–71 and 90–103 are sometimes shown below the main part of the table. They are not shown here.)

Questions

5 Name three elements in Group 1 of the periodic table.

6 Look at the elements in Period 3. From this period, name:
   a two metals
   b a non-metal.

B | The alkali metals

Find out about

- Group 1 metals
- reactions with water and chlorine
- similarities and differences between Group 1 elements

Cutting a lump of sodium to show a fresh, shiny surface of the metal.

highly flammable

corrosive

Questions

1 State three ways in which alkali metals are different from most other metals.
2 Write a word equation for each of the following reactions. Use the equation of lithium with water to help you.
   a Sodium with water.
   b Potassium with water.

The metals in Group 1 of the periodic table are very reactive. They include lithium (Li), sodium (Na), and potassium (K).

Chemists call these elements the **alkali metals** because they react with water to form alkaline solutions. The alkali metals are also corrosive and highly flammable. For this reason forceps should be used when handling these metals and goggles must be worn.

Strange metals

Like most metals, the alkali metals conduct electricity but their other properties are unusual. Most metals are shiny, hard, and strong. The alkali metals can be cut with a knife. The newly cut surfaces are shiny but they **tarnish** quickly in the air and become dull.

Most metals are dense and have a high melting point. Again the alkali metals are odd: lithium, sodium, and potassium float on water and melt on very gentle heating.

Reactions with water

The reactions of alkali metals with water should be carried out behind a safety screen. The reactions can be violent and particles can be thrown out from the surface of the water.

If you drop a small piece of lithium into water it floats, fizzes gently, and disappears as it reacts and forms lithium hydroxide (LiOH). This dissolves, making the solution alkaline. If the gas is collected, a burning splint can be used to show that it is hydrogen.

$$\text{lithium} + \text{water} \longrightarrow \text{lithium hydroxide} + \text{hydrogen}$$

The reaction of sodium with water is more exciting. The sodium melts and skates around on the surface of the water. It fizzes as hydrogen is formed. Sodium forms an alkaline solution of sodium hydroxide (NaOH).

The reaction of potassium with water is very violent. The metal moves around the surface of the water quickly. The hydrogen given off catches fire at once. The result is an alkaline solution of potassium hydroxide (KOH).

Reactions with chlorine

Hot sodium burns in chlorine gas with a bright yellow flame. It produces clouds of sodium chloride crystals (NaCl). This is everyday table salt, used for flavouring food.

The other alkali metals react with chlorine in a similar way. Lithium produces lithium chloride (LiCl). Potassium produces potassium chloride (KCl). Like everyday salt, these compounds are also colourless, **crystalline** solids that dissolve in water.

Chemists use the term **salt** to cover all the compounds of metals with non-metals. So the chlorides of lithium, sodium, and potassium are all salts.

Trends

The alkali metals are all very similar, but they are not identical. Their reactions with water show that there is a **trend** in their **chemical properties**. The alkali metals increase in reactivity down the group from lithium to sodium to potassium. There are also trends in their **physical properties**.

Sodium burning in chlorine gas.

	Melting point (°C)	Boiling point (°C)	Density (g/cm^3)
Lithium, Li	181	1342	0.53
Sodium, Na	98	883	0.97
Potassium, K	63	760	0.86

Physical properties of lithium, sodium, and potassium.

Lithium, sodium, and potassium all have a low density. The melting points and boiling points decrease down the group.

Summary box
- Group 1 metals include lithium, sodium, and potassium.
- Group 1 metals react vigorously with water and chlorine.
- The reactivity of the metals increases down the group.

Questions

3 Put the alkali metals (Li, K, Na) in order of reactivity from the least to the most reactive.

4 Rubidium is an alkali metal below potassium in the periodic table. Predict these properties of rubidium (Rb):
  a Its melting point.
  b What happens to a freshly cut surface of the metal in the air?
  c What happens if you drop a small piece of rubidium onto water?

5 Predict the formula of rubidium hydroxide.

C. Chemical equations

Find out about
- chemical symbols
- formulae

Equations are important because they show what goes in (reactants) and what comes out (products) of a reaction.

Chemical models

In a **chemical change** there is no change in mass because the number of each type of atom stays the same. The atoms rearrange, but no new ones appear and no atoms are destroyed.

Hydrogen burns in oxygen to form **molecules** of water. This reaction can be summarised as a word equation:

$$\text{hydrogen} + \text{oxygen} \longrightarrow \text{water}$$

The models in the diagram below show how the atoms are rearranged.

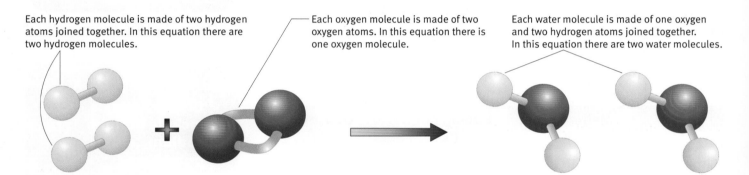

Each hydrogen molecule is made of two hydrogen atoms joined together. In this equation there are two hydrogen molecules.

Each oxygen molecule is made of two oxygen atoms. In this equation there is one oxygen molecule.

Each water molecule is made of one oxygen and two hydrogen atoms joined together. In this equation there are two water molecules.

There are equal numbers of hydrogen atoms and oxygen atoms on each side of the arrow.

Chemical symbols

Using models for every reaction would be very slow. Instead, chemists write symbol equations to show the numbers and arrangements of the atoms in the reactants and the products.

When written in symbols, the equation in the diagram above becomes:

$$2H_2 + O_2 \longrightarrow 2H_2O$$

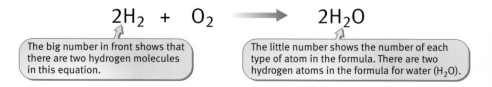

The big number in front shows that there are two hydrogen molecules in this equation.

The little number shows the number of each type of atom in the formula. There are two hydrogen atoms in the formula for water (H_2O).

The chemicals on the left of the arrow are the reactants and the ones on the right are the products.

Formulae

When you are using equations, it is important to use the correct chemical **formulae** for the reactants and products. Chemists have worked these out by experiment and you can look them up in data tables. The formula for each element or compound shows the number of each type of atom present. Chemists do not use a number one in a formula.

H_2O is the formula of a water molecule. It is made up of two hydrogen atoms and one oxygen atom.

Not all elements and compounds consist of molecules. For all metals, and the few non-metals that are not molecular (C, Si), you just write the symbol for a single atom.

Compounds of metals with non-metals are also not molecular. For these compounds you use the simplest formula for the compound, such as NaCl (sodium chloride), NaOH (sodium hydroxide), and $CaCO_3$ (calcium carbonate).

State symbols

State symbols are used to show whether the chemicals in the equation are solid (s), liquid (l), gas (g), or dissolved in water (aq).

Example: hydrogen + oxygen ⟶ water
$2H_2(g) + O_2(g) \rightarrow 2H_2O(l)$

Hydrogen and oxygen are gases and the water is a liquid.

> **Summary box**
> - In an equation, the big numbers are the numbers of each molecule present.
> - The small numbers are the numbers of atoms in the formula.
> - (s), (l), (g), and (aq) are state symbols.

Water is present on Earth as ice (s), water (l), and water vapour (g).

Questions

Look at this equation:
$Na(s) + H_2O(l) \rightarrow NaOH(aq) + H_2(g)$

1. In this equation, in what state is:
    a. sodium?
    b. water?
    c. sodium hydroxide?
    d. hydrogen?

2. How many oxygen atoms are there in the sodium hydroxide formula?

3. How many water molecules are there in the equation?

4. Write down the formula of:
    a. the reactants
    b. the products.

D The halogens

Find out about

- Group 7 elements
- halogen molecules
- similarities and differences between Group 7 elements

Dangerous elements

The elements in Group 7 of the periodic table are all very reactive non-metals. They include chlorine, bromine, and iodine. Group 7 elements are also known as the **halogens**.

The halogens are interesting because of their vigorous chemistry. As elements they are hazardous because they are so reactive. When working with halogens you should take the following safety precautions:

- Wear eye protection.
- Use chlorine and bromine in a fume cupboard because they are toxic gases.
- Wear chemical-resistant gloves when using liquid bromine, which is corrosive.

Halogen properties

Crystals of iodine.

- dense, yellow-green gas
- smelly and poisonous
- melting point −101 °C
- boiling point −35 °C

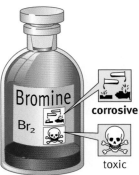

- deep-red liquid with red–brown vapour
- smelly and poisonous
- melting point −7 °C
- boiling point 59 °C

- grey solid with purple vapour
- smelly and harmful
- melting point 114 °C
- boiling point 184 °C

The halogens are made up of molecules. They each consist of **diatomic** molecules with the atoms joined in pairs: Cl_2, Br_2, and I_2. They have low melting and boiling points, which increase down the group.

All the halogens can harm living things. They can all kill bacteria. Chlorine is used to disinfect swimming pools. Iodine solution can be used on cuts and wounds to prevent infection.

Questions

1 Give one safety precaution you should take when working with bromine.

2 Fluorine is a halogen above chlorine in the periodic table. Would you expect its boiling point to be higher or lower than the boiling point of chlorine?

Reactions with metals

The halogens will react with alkali metals. These reactions show that the halogens get less reactive as you move down the group. The reactions of halogens with iron show the same trend or pattern.

- Hot iron glows brightly in chlorine gas. The product is iron chloride, which is a rust-brown solid.
- Iron glows less brightly when heated in bromine vapour.
- Iron reacts only slowly with iodine. Iron does not even glow when heated in iodine vapour.

Displacement reactions

The trend in reactivity of the halogens is also shown by **displacement reactions**. For example, if a pale green solution of chlorine is added to a colourless solution of sodium bromide, the solution immediately turns red because bromine has been formed. Chlorine is more reactive than bromine, so it displaces bromine from its salt, sodium bromide.

chlorine + sodium bromide ⟶ sodium chloride + bromine

In the same way, a solution of bromine will displace iodine from sodium iodide.

Bromine and iodine will not displace chlorine from its salts because bromine and iodine are less reactive than chlorine.

Hot iron in a jar of chlorine gas.

Questions

3 Put the halogens in order of reactivity from most to least reactive.

4 When iron is heated with chlorine, iron chloride is made. Predict the name of the compound made by heating:
  a iron and bromine   b iron and iodine.

5 Fluorine is the first element in group 7. Predict what would happen if fluorine was passed over iron. Write a word equation for the reaction.

6 Write a word equation for the reaction between:
  a chlorine and sodium iodide
  b bromine and sodium iodide.

Summary box
- Chlorine, bromine, and iodine are elements in Goup 7 of the periodic table. They are also called halogens.
- Halogens consist of diatomic molecules.
- Halogens are very reactive. They get less reactive as you go down the group.

E The discovery of helium

Find out about
- flame colours
- line spectra
- discovery of new elements

A new burner for chemistry

Robert Bunsen invented the Bunsen burner in 1855. Existing burners at the time produced smoky yellow flames. Bunsen wanted something better.

The great advantage of Bunsen's burner is that it can be adjusted to give an almost invisible flame. Bunsen used his burner to blow glass. He noticed that whenever he held a glass tube in a colourless flame, the flame turned yellow.

Flame colours

Soon Bunsen was experimenting with different chemicals, which he held in the flame at the end of a platinum wire. He found that different chemicals produced characteristic **flame colours**, like in the picture.

Bunsen thought that this might lead to a new method of chemical analysis, but he soon realised that it seemed only to work for pure compounds. It was hard to make any sense of flames from mixtures. So he mentioned his problem to Gustav Kirchhoff, a professor of physics.

Line spectra

'My advice', said Kirchhoff, 'is to look not at the colour of the flames, but at their spectra.'

Kirchhoff built a spectroscope from a glass prism and two telescopes. Light from a flame entered through one telescope. It was split into a spectrum by the prism and then viewed with the second telescope.

Bunsen and Kirchhoff soon found that each element has a different spectrum when its light passes through a prism. Each spectrum consists of a set of lines. With their spectroscope, they were able to record the **line spectra** of many elements.

Using **spectroscopy**, Bunsen discovered two new elements. He based their names on the colours of their spectra. He called them caesium and rubidium, from the Latin for 'sky blue' and 'dark red'.

The bright red flame produced by lithium compounds.

Robert Bunsen (1811–1899), who discovered the flame colours of elements with the help of his new burner.

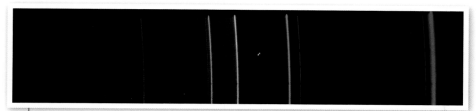

Seen using spectroscopy, cadmium's main lines are red, green, and blue (many fainter lines aren't visible in the photograph). The lines give cadmium its unique 'fingerprint'.

A Sun element

In 1868 there was a total eclipse of the Sun. Normally, the blinding light from the centre of the Sun makes it impossible to see the much fainter light from the hot gases around the edges. During an eclipse, the Moon hides the whole bright disc of the Sun but not the much fainter light from the hot gases around the edges. This makes it possible to study the light from these gases.

Pierre Janssen, a French astronomer, took very careful observations of the Sun's spectrum during the 1868 eclipse. In the spectrum of the light, he saw a yellow line where no yellow line was expected to be. He and an English astronomer, Joseph Lockyer, developed new methods to study the light from the Sun's gases. They worked independently, but both came to the same conclusion: there was a new element in the Sun, which they called helium. Because they had reproduced each other's findings, other scientists were confident that they had indeed found a new element.

During an eclipse it is possible to study the spectra of the light from the hot gases around the edges of the Sun. This is how helium was discovered in 1868.

WARNING!

Never look directly at the Sun, even during an eclipse. You can damage your eyes or even be blinded!

Questions

1. Look at the picture on page 54. What happens when lithium compounds are heated in a flame?
2. Name three elements that were discovered by spectroscopy.
3. In the story of Janssen and Lockyer, identify a statement that is data and a statement that is an explanation of the data.
4. Janssen and Lockyer discovered helium at about the same time. Explain why this meant that other scientists were more likely to accept their findings.

Summary box
- Many elements produce a colour in a flame.
- Line spectra were used to discover elements.

F Atomic structure

Find out about

- atomic theory
- the nuclear model of the atom
- protons, neutrons, and electrons

Part of the map of the London Underground.

1804
Dalton's solid atom.

1932
A model of the atom in which the nucleus is built up from neutrons and protons. The nucleus is surrounded by a cloud of electrons.
Not to scale

Atomic models from 1804 and 1932. In your study of chemistry, you will be using a model of the atom that dates back to 1932.

Question

1 What data could Dalton's theory of atomic structure not explain?

Atomic models

A picture, or model, of an atom can be used to understand how atoms join together to form compounds and how atoms rearrange during chemical reactions. Scientists use different models to solve different problems. There is not one model that is 'true'. Each model can represent only a part of what we know about atoms.

The London tube map is a kind of model. It is a very useful guide for getting from one station to another. It is 'true' in that it shows how the lines and stations connect, but it can't solve all of a traveller's problems. The map doesn't show how the tube stations relate to roads and buildings on the surface; for that you need a street map.

Dalton's atomic theory

The way we now think about atomic structure began with John Dalton. In 1803 Dalton was studying mixtures of gases. He did not just summarise the data he collected. He used his creativity to think up an explanation for what he found. In Dalton's theory, everything is made of atoms that cannot be broken down. He imagined atoms as solid spheres.

Dalton's theory explained how each element has its own kind of atom, and the atoms of different elements differ in mass. These ideas still apply today. Even so, Dalton's theory is limited. It cannot explain the pattern of elements in the periodic table. Nor can it explain how atoms join together in elements and compounds.

Inside an atom

Over time, more experiments were carried out. The model of the atom changed when new data was collected that could not be explained by Dalton's theory.

When scientists find new data and develop a new idea they write a paper for a scientific journal. Before a paper is published, other scientists read it. They check the way the scientist has done their work, the quality of their data, and how good their explanation is. This is called peer review.

If other scientists think the work is good then it is more likely to be published.

It is now an accepted theory that atoms are not solid spheres, but are made up of other, smaller particles.
- The middle of the atom is the **nucleus**. Nearly all the mass of the atom is here.
- The nucleus is made of **protons** and **neutrons**.
- Protons are positively charged.
- Neutrons are neutral. They have no charge.
- Around the nucleus are **electrons**, which are negatively charged.
- The mass of the electron is negligible. Its mass is so small that it can often be ignored.

The **proton number** is the number of protons in the atom. It is the same for all atoms of the same element. In an atom, the number of electrons is the *same* as the number of protons. This makes the atom neutral as the positive charge of the protons is balanced out by the negative electrons.

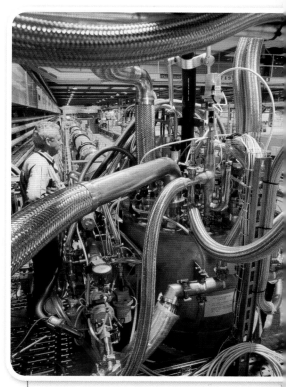

The Large Hadron Collider at CERN in Switzerland. Scientists are still working today to find out more about what an atom is made of. At CERN, particles are made to collide with each other at high speed, and the particles created in the collisons are studied.

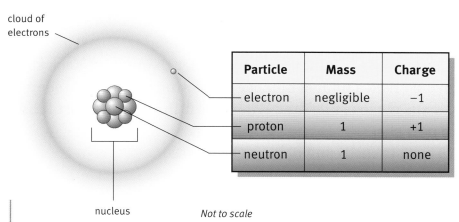

Particle	Mass	Charge
electron	negligible	−1
proton	1	+1
neutron	1	none

Not to scale

Atoms consist of three basic particles: protons, neutrons, and electrons.
The nucleus of an atom is very, very small.

Questions

2 What is the mass of an electron and what is its charge?

3 Which two particles are found in the nucleus?

4 Oxygen has proton number 8. Oxygen has how many:
   a protons?
   b electrons?

Summary box
- Atoms are made of protons, neutrons, and electrons.
- An atom has a tiny nucleus surrounded by a cloud of elctrons.

G Electrons in atoms

Find out about

- evidence for energy levels
- electrons in shells
- electron arrangements

Electrons in orbits

In 1913, the Danish scientist Niels Bohr came up with an explanation for the line spectra from atoms.

In Bohr's model the electrons in an atom circle the nucleus in an orbit. The electrons can move closer to or further away from the nucleus into new orbits, but only at particular distances from the nucleus. So for an electron to move it must 'jump' between orbits.

Bohr's idea was that heating atoms gives them energy. This energy makes the electrons move to a new orbit further from the nucleus. Soon these electrons jump back from outer orbits to inner orbits. As they do so they give out light energy of a particular colour. The colour depends on the size of the jump. The different colours are seen on the spectrum. Only certain energy jumps are possible so the spectrum consists of a series of lines.

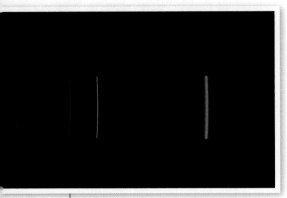

The line spectrum of hydrogen. Atomic theory can explain why this spectrum is a series of lines.

Electrons in shells

We now think of electrons in atoms as being arranged in different **shells** or **energy levels** rather than orbits. Each shell can only hold a limited number of electrons. The inside shell near the nucleus fills first. When it is full, electrons start filling the next shell.

- The first shell can hold two electrons.
- The second shell can hold eight electrons.
- Once the second shell is full, the third shell starts to fill.

If there are more electrons, they occupy further shells.

Question

1 Copy and complete this diagram of an atom.

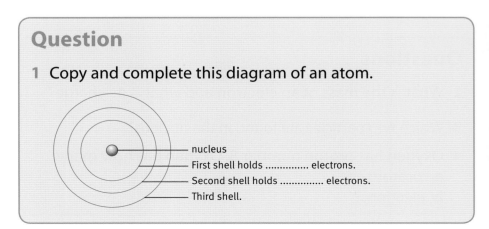

C4: CHEMICAL PATTERNS

Representing electron arrangements

Worked example

Sodium has 11 electrons. Draw a diagram to show how the electrons are arranged.

Step 1: Draw the nucleus. You can put in the symbol for the element if you wish.

Step 2: The first shell holds two electrons. Draw them as large dots.

Step 3: The second shells holds eight electrons. Draw them as large dots. There is one electron left.

Step 4: The last electron goes into the third shell.

This is a very useful diagram but there are times when a shorter version can be used. The arrangement of the electrons in a sodium atom can be written as 2.8.1.

This shows the same information as the diagram. Sodium has two electrons in the first shell, eight in the second shell, and one in the third shell.

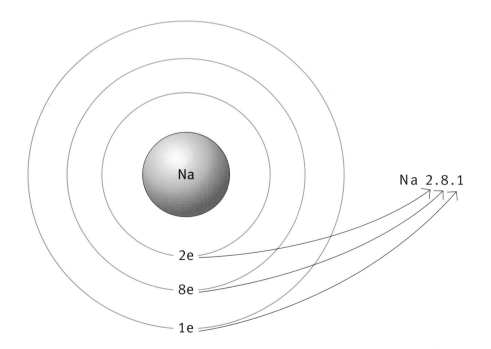

Summary box
- Electrons are arranged in shells or energy levels.
- The first shell can hold up to two electrons.
- The second shell can hold up to eight electrons.
- When the second shell is full, the third shell starts to fill.
- Line spectra provide evidence for energy levels.

Questions

2 Draw diagrams to show electrons in shells for these atoms:
  a beryllium (Be), 4 electrons
  b oxygen (O), 8 electrons
  c magnesium (Mg), proton number 12.

3 Show the electron arrangements of the following atoms using the shorthand version (eg, 2.8.1):
  a helium (He), 2 electrons
  b sulfur (S), 16 electrons
  c calcium (Ca), proton number 20.

G: ELECTRONS IN ATOMS

H Electrons and the periodic table

Find out about

- atomic structure and periods
- electron arrangements and groups

The periodic table then and now

Scientists discovered electrons in 1897, nearly 30 years after Mendeleev published his first periodic table. Mendeleev knew nothing about atomic structure, and he used the relative masses of atoms to put the elements in order.

A modern periodic table shows the elements in order of proton number, which is also the number of electrons in an atom. The 'shell model' of atomic structure helps to explain patterns in the periodic table.

Periods

The diagram below shows the connection between the horizontal rows of the periodic table and the structure of atoms. From one atom to the next, the proton number increases by one and the number of electrons increases by one. So the electron shells fill up progressively from one atom to the next.

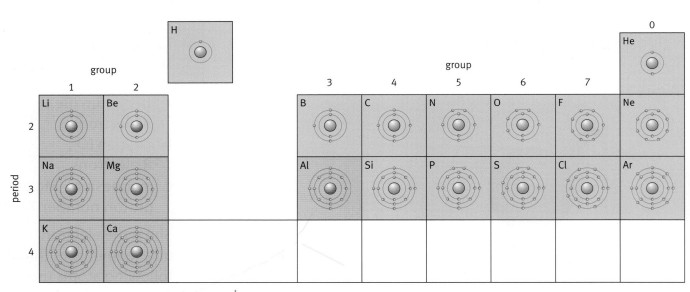

Electron arrangements for the first 20 elements in the periodic table.

In the first period, from hydrogen to helium, the first shell is being filled. The second shell fills across the second period, from lithium (2.1) to neon (2.8). Eight electrons go into the third shell, from sodium (2.8.1) to argon (2.8.8), and then the fourth shell starts to fill from potassium and calcium. After calcium, the arrangement of electrons becomes more complex.

Groups

For elements in Groups 1 to 7 the number of electrons in the outer shell of the element is the same as its group number.

Lithium, sodium, and potassium are in Group 1. They all have one electon in their outer shell.

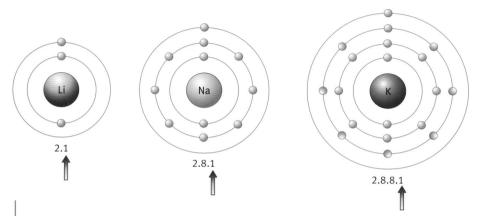

The electron arrangements of lithium, sodium and potassium. They all have one electron in their outer shell.

Lithium, sodium, and potassium have similar properties because they all have one electron in their outer shell.

Fluorine and chlorine are in Group 7. They have seven electrons in their outer shell.

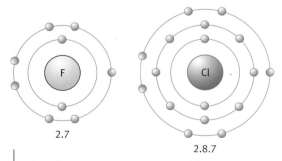

The electron arrangements of fluorine and chlorine.

Bromine and iodine have seven electrons in their outer shells too. The Group 7 elements have similar properties because they all have seven electrons in their outer shell.

The elements in Group 0 all have a full outer shell of electrons.

Summary box
- In a modern periodic table, atoms are arranged by proton number.
- Each electron shell fills up across a period in the periodic table.
- The number of electrons in the outer shell of an atom is the same as the group number.

Questions

1. In what order are the elements in the modern periodic table arranged?

2. Copy and complete:

   Each shell or energy level fills with _____ across a _____ in the periodic table.

3. What is the shorthand form of the electronic arrangement of:
   a potassium (K)?
   b argon (Ar)?
   c fluorine (F)?

4. Why do the elements in Group 7 have similar properties?

Salts

Find out about

- salts
- properties of salts
- electricity and salts

Why are salts so different from their elements?

Compounds of metals with non-metals are salts. Chemists can explain the differences between a salt and its elements by studying what happens to the atoms and molecules as they react. A good example is the reaction between two very reactive elements to make the everyday table salt you can safely sprinkle on food.

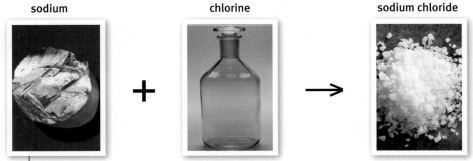

A chemical reaction in pictures: sodium and chlorine react to make sodium chloride.

White sodium chloride crystals are shaped like cubes. Sodium chloride is soluble in water.

Salts

Salts such as sodium chloride are crystalline. The crystals of sodium chloride are shaped like cubes. Salts have much higher melting and boiling points than chemicals such as chlorine and bromine, which are made up of small molecules.

Chemical	Formula	Melting point (°C)	Boiling point (°C)
sodium	Na	98	890
chlorine	Cl_2	−101	−34
sodium chloride	NaCl	808	1465
potassium	K	63	766
bromine	Br_2	−7	58
potassium bromide	KBr	730	1435

Crystals of the mineral pyrite. Pyrite consists of insoluble iron sulfide.

Sodium chloride is an example of a salt that is soluble in water. There are many other soluble salts. Some salts are insoluble in water. Iron sulfide, found in the mineral pyrite, is one example.

Molten salts and electricity

The apparatus on the right is used to investigate whether or not chemicals conduct electricity. The crucible contains some white powdered solid. This is zinc chloride.

At first no current flows. The solid does not conduct electricity. This is true of all compounds of metals with non-metals; they do not conduct electricity when solid.

Heating the crucible melts the zinc chloride. As soon as the compound is **molten** (melted), there is a reading on the meter. This shows that a current is flowing round the circuit. As a liquid, the compound is a conductor.

The electric current causes the compound to decompose chemically. There is bubbling around the positive electrode. The gas produced is chlorine. A shiny metal can be seen at the negative electrode. This is zinc.

So the electric current splits the compound into its elements: zinc and chlorine.

Salts in solution and electricity

Dissolved salts also conduct electricity. This can be studied using the apparatus shown on the right. There are changes at the electrodes when an electric current flows.

The presence of water has an effect on the chemicals produced when a salt solution conducts electricity. The products are not always the same as the elements in the compound.

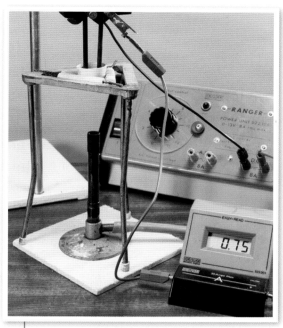

The crucible contains zinc chloride. The carbon rods dipping into the crucible are the electrodes. A current begins to flow in the circuit when the zinc chloride is hot enough to melt.

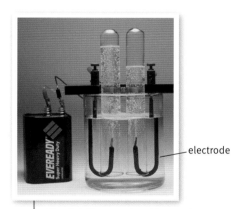

electrode

Electricity passing through a salt solution. Gases are being produced at the electrodes.

Questions

1 Make a table to compare the properties of sodium, chlorine, and sodium chloride. Write in your table the state at room temperature, colour, and hazards of each chemical.

2 Predict two properties of the salt sodium bromide.

Summary box

- Salts are compounds of a metal and a non-metal.
- Salts conduct electricity when molten or dissolved in water.

J — Ionic theory

Find out about
- ions
- ionic compounds
- explaining properties of salts

Electrolysis

An electric current can split a salt into its elements or other products if it is molten or dissolved in water. This is called **electrolysis**.

The discovery of electrolysis was very important in the history of chemistry. An English chemist, Humphry Davy, used electrolysis to isolate the elements potassium, sodium, barium, strontium, calcium, and magnesium for the first time.

Faraday's theory

Michael Faraday worked with Humphry Davy. In 1833 he began to study the effects of electricity on chemicals. He had to think creatively to come up with an explanation for his observations.

Faraday decided that compounds that can be decomposed by electrolysis must contain electrically charged particles. Since opposite electrical charges attract each other, he could imagine the negative electrode attracting positively charged particles and the positive electrode attracting negatively charged particles.

The charged particles move towards the electrodes. When they reach the electrodes, they turn back into atoms. Faraday called the charged particles **ions**.

Faraday's theory accounted for his observations and enabled him to make predictions. He published his work and gave lectures to share his ideas with other scientists.

Questions

1. What was Faraday's explanation for the effect of electricity on chemicals?
2. Why was it important for Faraday to communicate his ideas to other scientists?

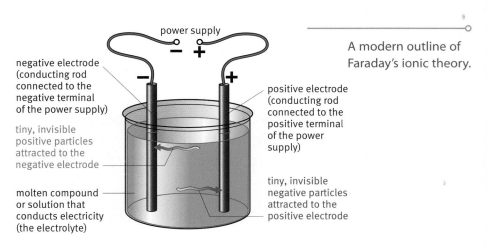

A modern outline of Faraday's ionic theory.

Explaining electrolysis

Chemists continue to use ideas about ions to explain electrolysis.

Salts such as sodium chloride consist of ions. Sodium chloride is made up of sodium ions and chloride ions. Sodium ions, Na^+, are positively charged. The chloride ions, Cl^-, carry a negative charge. These oppositely charged ions attract each other.

A crystal of sodium chloride consists of millions and millions of Na^+ and Cl^- ions closely packed together. In the solid, these ions cannot move towards the electrodes, and so the compound cannot conduct electricity.

The ions can move when sodium chloride is hot enough to melt or when it is dissolved in water. When the ions can move, the salt will conduct electricity.

Metal ions are positive and non-metal ions are usually negative. During electrolysis, the negative electrode attracts the **positive ions**. The positive electrode attracts the **negative ions**.

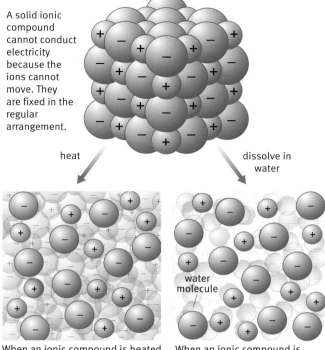

A solid ionic compound cannot conduct electricity because the ions cannot move. They are fixed in the regular arrangement.

When an ionic compound is heated strongly and melts, the ions can move around and the molten compound conducts electricity.

When an ionic compound is dissolved in water, it can conduct electricity because its ions can move among the water molecules.

Questions

3 What sort of ions (positive or negative) are usually formed by:
   a non-metals?
   b metals?

4 Why don't solid compounds made of ions conduct electricity?

5 What happens to the ions when an ionic solid dissolves in water? (Look at the diagram to help you.)

6 Why do dissolved ionic compounds conduct electricity?

Summary box
- Salts can be split up by electricity if they are dissolved in water or molten. This is evidence that they are made of ions.
- Ions are particles with a positive or negative charge.

K | Ionic theory and atomic structure

Find out about

- atoms and ions
- electron configurations of ions

Atoms into ions

Faraday could not explain how atoms turn into ions because he was working long before anyone knew anything about the details of atomic structure. Today, chemists can use the shell model for electrons in atoms to show how atoms become electrically charged.

Metals form positive ions

The metals on the left-hand side of the periodic table have only a few electrons in their outer shells. They form ions by losing these electrons. This leaves more positively charged protons than negatively charged electrons. So the ions are positively charged. All the metals in Group 1 have one electron in the outer shell. They all lose that electron to form ions with a 1+ charge, for example, Li^+, Na^+, and K^+.

sodium atom, Na 2.8.1
number of electrons: 11
number of protons: 11
overall charge: 0

loses the electron from the outer shell

sodium ion, Na^+ 2.8.
number of electrons: 10
number of protons: 11
overall charge: +1

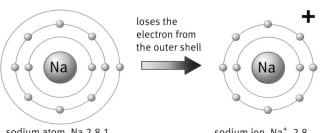

A sodium atom turns into a positive ion when it loses a negatively charged electron.

Questions

1 Lithium is in Group 1 of the periodic table. It has a proton number of 3.
   a How many electrons does a lithium atom have?
   b Draw a diagram to show the electron arrangement of a lithium atom.
   c How many electrons are in the outer shell of a lithium atom?
   d What is the charge on a lithium ion?
   e Draw a diagram to show the electron arrangement of a lithium ion.

2 a A potassium atom has 19 electrons. Write down a shorthand version of its electron arrangement.
   b Write down the electron arrangement of a potassium ion and state its charge.

3 Caesium is in Group 1. Predict the charge on a caesium ion.

Non-metals form negative ions

The non-metals on the right-hand side of the periodic table have nearly full shells of electrons. They form ions by gaining extra electrons to fill up their shells. This means they have more negatively charged electrons than positively charged protons. So the ions are negatively charged.

All the halogens in Group 7 have seven electrons in their outer shells. The outer shells can hold up to eight electrons. They all gain an extra electron to form ions with a −1 charge, for example, Cl^-, Br^-, and I^-.

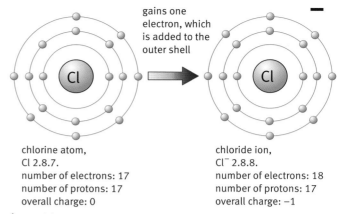

chlorine atom, Cl 2.8.7.
number of electrons: 17
number of protons: 17
overall charge: 0

chloride ion, Cl^- 2.8.8.
number of electrons: 18
number of protons: 17
overall charge: −1

A chlorine atom turns into a negative ion by gaining an extra negatively charged electron.

Ions into atoms

Electrolysis turns ions back into atoms. Metal ions are positively charged, so they are attracted to the negative electrode. It is a flow of electrons from the battery into this electrode that makes it negative. Positive metal ions gain electrons from the negative electrode and turn back into atoms.

Non-metal ions are negatively charged, so they are attracted to the positive electrode. This electrode is positive because electrons flow out of it to the battery. Negative ions give up electrons to the positive electrode and turn back into atoms.

The elements and atoms of Group 7 have names ending in -ine, for example, chlorine. The ions of Group 7 have names ending in -ide, for example, chloride.

Question

3 Fluorine is in Group 7 of the periodic table. It has a proton number of 9.
   a How many electrons does a fluorine atom have?
   b Draw a diagram to show the electron arrangement of a fluorine atom.
   c How many electrons are in the outer shell of a fluorine atom?
   d What will the charge be on a fluoride ion?
   e Draw a diagram to show the electron arrangement of a fluoride ion.

Summary box

✓ Group 1 metals form positive ions by losing an electron.
✓ Group 7 elements form negative ions by gaining an electron.

L Chemical species

Find out about

- atoms, molecules, and ions
- chemical species

In this module you have met the idea that the same element can take different chemical forms with distinct properties. Chemists describe these different forms as **chemical species**.

Species of chlorine

Chlorine has three simple species: atom, molecule, and ion. Each of these species of chlorine has distinct properties.

Chlorine atoms (Cl) do not normally exist in a free state. They rapidly pair up to form chlorine molecules (Cl_2).

Ultraviolet radiation can split chlorine (and chlorine compounds) into atoms. This is what happens to CFCs such as CCl_3F when they get into the upper atmosphere.

When the molecules break up into atoms, the highly reactive free chlorine atoms rapidly destroy ozone. This creates the so-called 'hole' in the ozone layer.

Chlorine gas at room temperature consists of chlorine molecules (Cl_2). These are very reactive, as illustrated by the reactions of the halogens with iron described in Section D. The chlorine molecules can damage human tissues, so the gas is given the label 'toxic'.

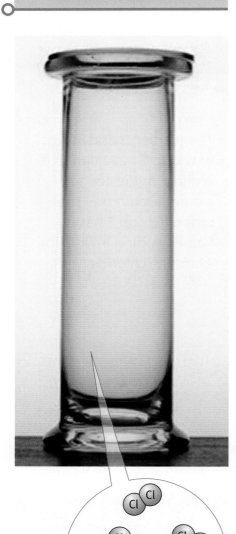

Chlorine gas consists of diatomic chlorine molecules. Chlorine molecules are chemically very reactive.

Polar stratospheric cloud formations over the Arctic. This is where CFCs break down and can destroy ozone.

Chloride

Chloride ions are quite different. They occur in compounds such as sodium chloride, or table salt and magnesium chloride. Chloride ions in salts are essential to life and occur in all living tissues. Chloride ions are chemically active in many ways, but they are not as reactive and harmful as the atoms or molecules of the element.

Species of sodium

There are only two species of sodium: atom and ion. The atoms in sodium metal are chemically very active. In the presence of other chemicals, the sodium atoms react to produce compounds containing sodium ions.

Sodium combines with chlorine to produce the ionic compound sodium chloride. This is made up of two chemical species: Na^+ and Cl^-. These two ions are quite unreactive. Sodium chloride is soluble in water, but its solution is neutral. Water does not react with the ions.

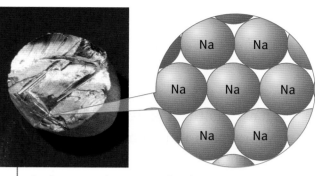

Sodium metal consists of sodium atoms. Sodium atoms are chemically very reactive.

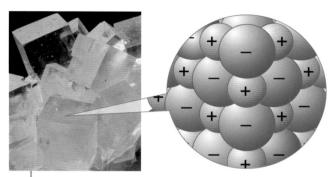

Sodium chloride consists of sodium ions and chloride ions. These ions are not very reactive.

Questions

1 Name three different types of chemical species.
2 Use the idea of chemical species to explain why the properties of sodium chloride are very different from the properties of its elements sodium and chlorine.

Summary box

- Chlorine atoms and molecules are very reactive, but chloride ions are not.
- Sodium atoms are very reactive, but sodium ions are not.

Science Explanations

Chemists have identified patterns in the world and come up with theories to explain how roughly 100 elements can combine to make such a huge variety of chemical compounds.

You should know:

- that the chemists' model of the atom has a tiny central nucleus, containing protons and neutrons, and that the nucleus is surrounded by electrons
- that the number of electrons in an atom is equal to the number of protons
- that electrons are arranged in shells (energy levels) with the innermost electron shell, which has the lowest energy, filling first until full then the next shell starts to fill, and so on
- that the periodic table consists of elements lined up in order of their proton numbers, so that there are repeating patterns across each row (period) in the table
- that each column in the periodic table consists of a group of related elements with similar chemistries (properties and behaviour) because they have the same number of electrons in their outer shells
- that Group 1 elements are called alkali metals and they react with moist air, water, and chlorine, becoming more reactive down the group
- that Group 7 elements are called halogens and they are made up of molecules made of two atoms
- that the different halogens become less reactive down the group
- that chemists use word equations and symbol equations to describe reactions
- why safety precautions are important when working with hazardous chemicals such as Group 1 metals, alkalis, and the halogens
- how ionic compounds form when metals react with non-metals such that metal atoms lose electrons while the non-metal atoms gain electrons
- how a regular lattice of ions gives rise to the shape of the crystals of an ionic compound
- that the properties of an ionic compound are the properties of its ions, which behave in a different way to the atoms or molecules of its elements
- that ionic compounds conduct electricity when molten or when dissolved in water, because the ions are charged and they are able to move around in the liquid.

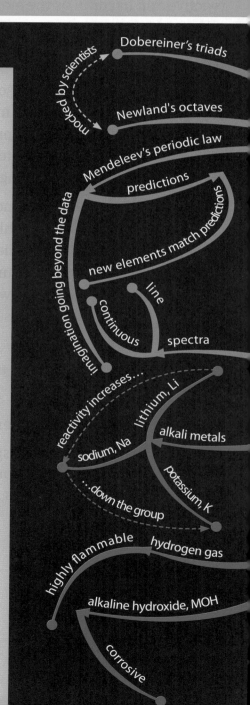

C4: CHEMICAL PATTERNS

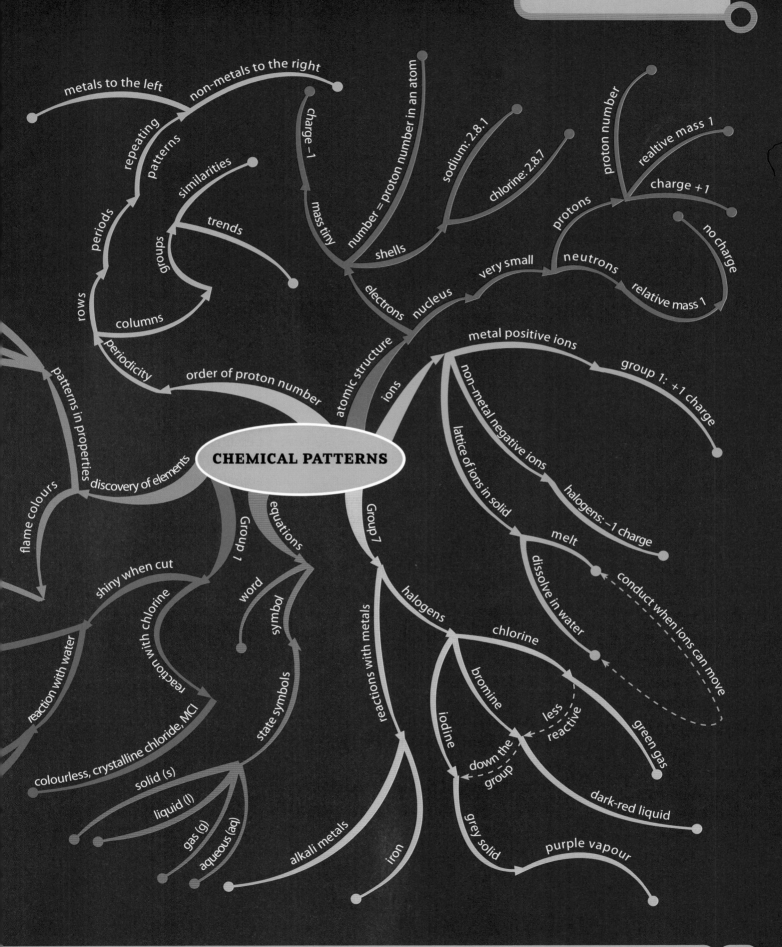

Ideas about Science

Scientific explanations are based on data but they go beyond the data. An explanation has to be thought up creatively to account for the data. A new explanation may explain a range of observations or events not previously thought to be linked. The explanation should also allow predictions to be made.

In the context of the discovery of new elements and the development of the periodic table you should be able to:

- identify statements that report information collected (data) and statements of explanatory ideas (hypotheses, explanations, and theories) in an account of scientific work
- recognise that an explanation may be incorrect even if the data is correct
- identify where creative thinking has been used in the development of an explanation, such as Mendeleev's insight that he had to leave gaps for undiscovered elements
- recognise data or observations that are accounted for by (or conflict with) an explanation, such as the data from spectra that could be explained by the existence of new elements
- give good reasons for accepting or rejecting a proposed scientific explanation, such as the way that the shell model of atomic structure accounts for the arrangement of elements in the periodic table
- identify which scientific explanation best explains the data, where more than one explanation is proposed, for instance, Mendeleev's use of his periodic table to predict the existence of unknown elements
- understand that when a prediction agrees with an observation this increases confidence in the explanation on which the prediction is based, but does not prove it is correct; for example, Mendeleev correctly predicted the properties of missing elements in his periodic table

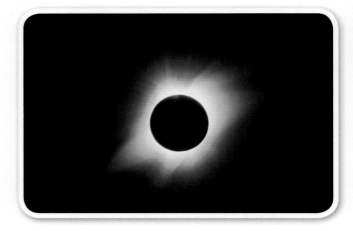

- understand that when a prediction disagrees with an observation this indicates that one or the other is wrong and decreases confidence in the explanation on which the prediction is based.

Scientists report their claims to other scientists through conferences and journals. Scientific claims are only accepted once they have been evaluated by other scientists. Scientists are usually sceptical about claims that cannot be reproduced by anyone else and about unexpected findings until they have been repeated (by themselves) or reproduced (by someone else).

You should be able to:

- broadly outline the peer-review process, in which new scientific claims are evaluated by other scientists
- recognise that there is less confidence in a new scientific claim that has not yet been evaluated by the scientific community than a well-established claim; for instance, early suggestions of connections between the chemical properties of elements and their relative atomic mass were not thought to be reliable and were not accepted
- identify the fact that a finding has not been reproduced by another scientist as a reason for questioning a scientific claim.

Review Questions

C4: CHEMICAL PATTERNS

  A teacher reacts sodium with chlorine. The product is sodium chloride, NaCl.
  a The sodium bottle has two hazard symbols.

  i Give the meaning of each symbol.
  ii List two safety precautions the teacher must take when using sodium. Link each precaution to a hazard.
  b Write a word equation for the reaction of sodium with chlorine.
  c Sodium chloride is made up of two types of ion.
   i A sodium atom has 11 protons. How many electrons does it have?
   ii Draw a diagram to show the electron arrangement of a sodium atom.
   iii What is the charge on a sodium ion?
   iv A chlorine atom gains an electron to become an ion. What is the charge on a chloride ion?

  The table shows part of the periodic table, with only a few symbols included.

group\period	1	2	3	4	5	6	7	0
1								He
2						O	F	
3	Na	Mg	Al	Si		S	Cl	
4				Ge			Br	
5								
6	Cs							

  a Using only the elements in the table, write down symbols for:
   i a metal that floats on water
   ii an element with similar properties to silicon (Si)
   iii two metals from Period 3
   iv three elements that have molecules made up of two atoms
   v the most reactive metal.
  b Chlorine is a gas at room temperature and bromine is a liquid.
   i Describe the trend in boiling points of the halogens.
   ii The symbol for fluorine is F. Predict the state of fluorine at room temperature. Give a reason for your prediction.

  The halogens react with iron.
  a Hot iron glows brightly in chlorine gas. Iron does not glow when heated with iodine vapour. Predict what happens when iron is heated with bromine vapour. Explain your answer.
  b The balanced equation for the reaction of iron with bromine is:
  $$2Fe + 3Br_2 \longrightarrow 2FeBr_3$$
   i What is the name of the product of this reaction?
   ii How many bromine molecules are there in this equation?
   iii How many bromine atoms are there in the formula of the product?

  Chlorine reacts with a solution of potassium bromide to produce bromine and potassium chloride solution.
  a Write a word equation for this reaction.
  b What type of reaction is this?
  c Bromine does not react with potassium chloride. Why not?
  d What chemicals could be mixed to show that iodine is less reactive than bromine?

P4 Explaining motion

Why study motion?

Humans have always been interested in how things move and why they move the way they do. Moving things play a big part in our everyday lives so we really need to be able to explain and predict how objects move.

What you already know

- Speed is calculated by the distance covered divided by time.
- An unbalanced force changes the motion of an object.
- The weight of an object is due to the gravitational force between the Earth and the object.
- The air resistance on a moving object depends on its shape and its speed.

Find out about

- forces always arising from an interaction between two objects
- friction and reaction of surfaces
- instantaneous and average speed, velocity and acceleration
- how the momentum of an object changes when a force acts on it
- everyday examples of motion
- gravitational potential energy and kinetic energy.

The Science

Every example of motion we observe can be explained by a few simple rules (or laws) that apply to everything. These laws are so exact and precise that they can be used to predict the motion of an object very accurately. A key idea is the force acting on an object and how an unbalanced force changes the motion of the object.

Ideas about Science

Understanding forces and motion has enabled scientists and engineers to design and build more efficient cars and trains, to develop aircraft, and to fly spacecraft with enormous accuracy to the furthest reaches of the Solar System.

A Forces and interactions

Find out about

- how forces arise when two objects interact
- contact and action-at-a-distance forces

To start something moving we have to push or pull it. A force is needed. And to stop something moving it needs a force acting against it.

Forces always come in pairs

Look at the photograph of a firework rocket exploding. The burning sparks move out from the centre where the chemicals in the rocket have exploded. Notice that the starburst is symmetrical. For every moving spark another spark moves in exactly the opposite direction. Forces always come in pairs.

An interaction pair of forces

The diagram shows Sophie and Sam standing in the centre of an ice rink. When Sophie gives Sam a push both of them start to move. Sophie can't make Sam move without moving backwards herself, however hard she tries. This means that when Sophie exerts a force on Sam by pushing him Sam exerts an equal and opposite force on Sophie – not by pushing but just by being there as something to push against. It's the same for Sam.

This works with pulling forces too. When Sophie and Sam stand a distance apart, holding a rope, and one of them pulls, both of them start to move towards each other. Every force arises from an interaction between a pair of objects.

In this example one force acts on Sam and the other force acts on Sophie.

The chemical reaction inside the firework produces forces that send the burning fragments out equally in all directions, producing a sphere of sports.

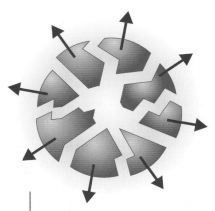

An exploding firework rocket. For every moving spark, there is another spark moving in exactly the opposite direction.

Forces always come in pairs. When Sophie pushes Sam, she gets pushed too.

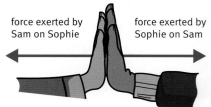

These forces are equal and opposite. Contact forces always come in pairs.

76

Contact forces

When forces are caused by two objects touching, we call them contact forces. Contact forces exist only while the objects are touching. As soon as the objects separate, the forces stop. But the objects may keep moving.

Action-at-a-distance forces

Magnetism is an example of a force that acts at a distance. A magnet can attract without touching. Two magnets can also repel without touching.

Gravity is another example of action at a distance. The apple in the diagram falls downwards because it is pulled towards the centre of the Earth. But gravity is an attraction between two objects. The apple exerts an equal and opposite pull on the Earth! This is not noticed because the Earth is so massive.

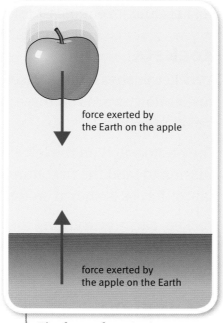

The force of gravity is an example of an interaction pair.

Forces in diagrams

Forces are represented by arrows.

The direction of the arrow shows the direction of the force.
The length of the arrow represents the size of the force.

⟶ 10 N to the right

⟵ 5 N to the left

Both the fridge magnet and the fridge experience a force. Action-at-a-distance forces always come in pairs.

Questions

1. Complete these examples of interaction pairs of forces.
   a Sophie pushes Sam and Sam …
   b A horse pulls on a rope and the …
   c Jack pushes on a wall and the …
   d A train exerts a force on the buffers and the …

2. List three examples of interaction pairs of forces on these pages.

3. What three things are always true about interaction pairs?

Summary box
- ✓ **Forces always come in interaction pairs**
- ✓ **The forces in an interaction pair are:**
  - **equal in size**
  - **opposite in direction**
  - **act on different objects.**

B Getting moving

Find out about

- the forces that enable people and vehicles to get moving
- rockets and jet engines

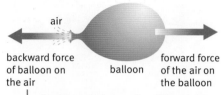

An interaction pair of forces.

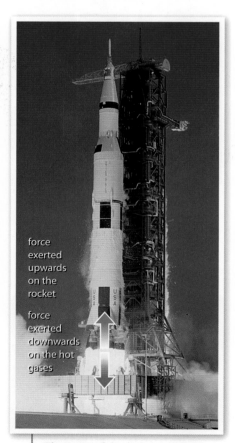

The start of the longest journey humans have made so far – to the Moon. A huge force was needed to push the rocket upwards.

An interaction pair of forces is needed to get something moving. Here are some examples.

Flying balloon

When you let go of an inflated balloon it flies around. The diagram shows the interaction pair of forces that make it move. Rockets and jet engines work in a similar way.

Rockets

A rocket pushes out hot gases as its fuel burns. The rocket pushes downwards on these gases. The escaping gases push the rocket upwards.

The photograph on the left shows the launch of the *Apollo 11* mission to land the first humans on the Moon. The interaction pair of forces is shown on the photo.

Rockets carry everything they need to make the burning gases they push out. This means that they can work anywhere, including empty space.

Jet engines

Air is pushed out at high speed from the back of a jet engine. The force of the air on the engine pushes it forward. Jet engines need to draw air in, so they cannot work in space.

Questions

1 Copy and complete the table.

	Interaction pairs of forces	
	Force 1	Force 2
Balloon	force of the balloon on the air	
Jet engine		
Rocket		

2 Explain why:
   a jet engines cannot travel in space
   b rockets can travel in space.

How does a car start moving?

To make a car move, the wheels turn. This causes a forward force on the car. To understand how, think first about a car trying to start on ice. If the ice is very slippery, the wheel will just spin. The car will not move at all. The spinning wheel produces no forward force on the car. Now imagine a car on a muddy track. The rally car below is throwing up a shower of mud as it tries to get going.

You can see that the wheel is causing a backwards force on the ground surface. This makes the mud fly backwards. Mud moves when the force is small. The other force of the interaction pair is the forward force on the car. It is equal in size, so it is too small to get the car moving.

Now imagine a good surface and a good tyre, which does not slip. The wheel turns. It pushes back on the road. The wheel does not slip; it exerts a very large force backwards on the road. The other force of the interaction pair is the same size. This large forward force gets the car moving.

As it rotates, the wheel exerts a force backwards on the ground.

Question

3 Add to your table for question 1 the interaction pair of forces:
   a on a car wheel
   b on a boat propeller (a boat propeller pushes water backwards when it spins round).

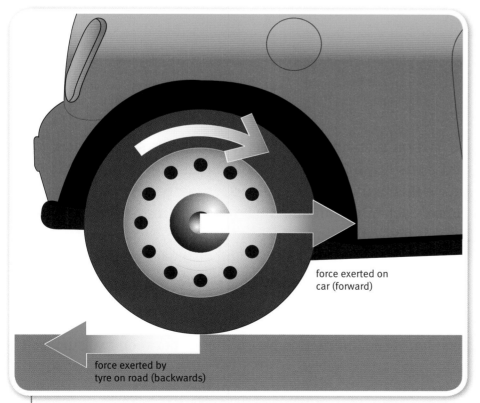

force exerted on car (forward)

force exerted by tyre on road (backwards)

The tyre grips the road and does not spin. The road exerts a large forward force on the axle. This pushes the car forward.

Summary box
- **An interaction pair of forces is needed to start an object moving.**

C Friction

Find out about
- friction and what causes it
- how to add the forces acting on an object

Friction often seems a nuisance. But without it, you could not start moving, car wheels would just spin.

How much friction?

Jeff is trying to push a large box along a level floor.

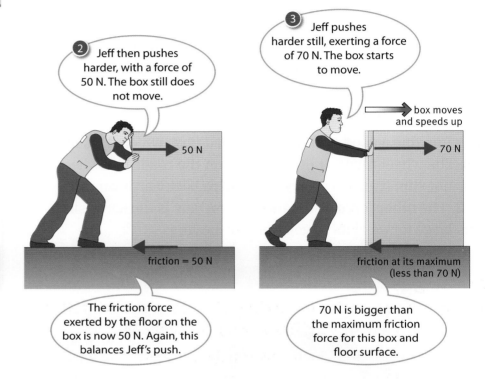

1. Jeff pushes the box with a force of 25 N, to try to slide it along. It does not move.

   friction = 25 N

   The friction force exerted by the floor on the box is 25 N. This exactly balances Jeff's push.

2. Jeff then pushes harder, with a force of 50 N. The box still does not move.

   friction = 50 N

   The friction force exerted by the floor on the box is now 50 N. Again, this balances Jeff's push.

3. Jeff pushes harder still, exerting a force of 70 N. The box starts to move.

   box moves and speeds up

   friction at its maximum (less than 70 N)

   70 N is bigger than the maximum friction force for this box and floor surface.

So friction is an unusual force. It changes its size to match Jeff's pushing force – up to a certain limit. If Jeff's push is below this limit, friction matches it exactly. The two forces cancel each other out and the box doesn't move. But if Jeff's force is above this limit, the box will move. So what determines this limit? It depends on the weight of the box, and on the roughness of the two surfaces in contact.

What is friction?

Friction is the interaction pair of forces between two surfaces when they slide over each other – or when they are pushed to try to make them slide over each other.

The bumps and hollows on each surface cause a sideways force, as one object slides (or tries to slide) across the other.

What causes friction?

Friction is caused by the roughness of the surfaces. The diagram shows that even surfaces that seem smooth have quite large humps and hollows if you look at them under a microscope. Some of the bumps on one surface will fit into hollows on the other. As one slides across the other, these bumps collide. This causes a sideways force that resists the sliding.

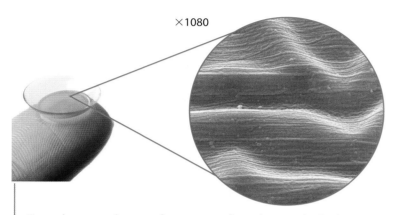

Even the smoothest surfaces are really quite rough. At the microscopic level they have humps and hollows. This photograph shows the surface of a contact lens magnified 1080 times.

Adding forces

If there is a force acting on an object, but it is not moving, then there must be another force balancing the first one.

If all the forces acting on an object balance each other they add to give a resultant force of zero. If all the forces on an object are added together and they don't balance, then they give a resultant force.

The diagrams show how to add forces to find the resultant force. You must take the direction of each force into account.

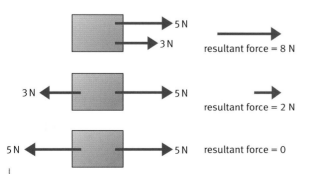

To find the resultant force you add the forces. You must take account of their directions.

Questions

1 List three everyday situations in which:
   a we try to reduce friction
   b we try to make friction as large as possible.

2 Explain why it is easier to push a box across the floor when it is empty than when it is full.

3 Work out the resultant force of these forces:

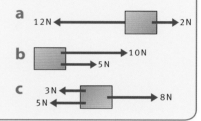

Summary box

✓ Friction is the interaction pair of forces between two surfaces that try to slide over each other.
✓ Forces can be added to give a resultant force.

D | Vertical forces

Find out about

- how a surface exerts a reaction force on any object that presses on it
- the forces acting on a falling object

When you drop a tennis ball, it immediately starts to move downwards. The force acting on the tennis ball is gravity. It is the downward pull of the Earth.

Balanced forces

But if you put a tennis ball on a table, it does not fall. The force of gravity has not suddenly stopped or been switched off. There must be another force that balances it out. The only thing that can be causing this is the table. The table must exert an upward force on the ball that balances the downward force of gravity.

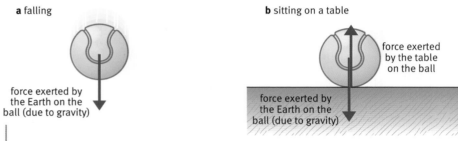

The forces acting on a tennis ball **a** falling and **b** sitting on a table.

The two forces in diagram **b** are balanced. They are not an interaction pair of forces because they both act on the same object – the tennis ball.

How can a table exert a force?

Look at the diagram of the school bag, sitting on the foam cushion. The bag presses down on the foam, squashing it. Because foam is springy, it pushes upwards on the bag, just like a spring. Like a spring, the more it is squeezed, the harder it pushes back. So the bag sinks into the foam until the push of the foam on the bag exactly balances the downward pull of gravity on it.

The same thing happens when the bag sits on a table top. You can't see that the table top is being squashed because it is squashed by such a tiny amount. We call this upward force that a surface exerts when something presses on it the **reaction** of the surface.

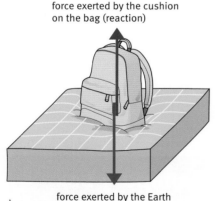

The bag squeezes the foam until the upward force of the springy foam on the bag exactly balances the downward force of gravity on the bag.

The size of the reaction force depends on the downward force that is causing it. The surface squashes just enough for the reaction force to balance the force of gravity. The resultant force on the object is zero.

If the downward force exerted on a table is bigger than it can take, the table top will break!

Walls can push too!

The diagram on the right shows what happens when Deborah pushes on the wall. The wall is compressed. Like a spring, it pushes back with an equal force on Deborah's hands. She immediately starts to move backwards.

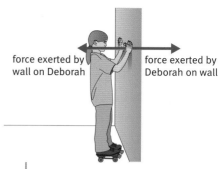

Deborah pushes against the wall. The wall pushes on Deborah. She starts moving backwards.

Freefall

The multi-flash photograph on the right shows that the tennis ball picks up speed steadily from the moment it is released. As the ball falls, it moves further between one flash and the next. Its speed increases. It accelerates.

But when we drop an object that is light relative to its size, like a paper muffin case, it speeds up at first, then falls at a steady speed. The reason is that **air resistance** has a bigger effect on it. Air resistance is a force that arises when anything moves through the air. The faster an object moves, the bigger the air resistance force on it becomes. The falling muffin case quickly reaches the speed at which the air resistance force on it balances the force of gravity. It then continues to fall at a steady speed.

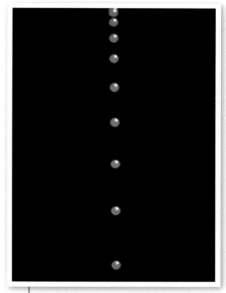

Freefall – a falling ball gets steadily faster as it falls.

Questions

1. Draw a diagram showing the balanced forces when a ball is on a table.
2. Draw a diagram showing the interaction pair of forces when a ball hits a wall.
3. What happens to a 'hard' surface when something sits on it?

Summary box
- The reaction force is the force against an object pressing on a surface. It balances the pressing force – up to a limit.
- The forces on an object in freefall are gravity and air resistance.

E. Describing motion

Find out about

- how to calculate the speed of a moving object
- how to calculate the acceleration of a moving object

There are some links between forces and motion. To explore these more fully, we need to be able to describe the motion of an object more clearly.

Speed

The diagram shows a 6 km journey. If it takes 1 hour to complete we can say the speed is 6 km per hour.

To find the speed of an object, we measure the time it takes to travel a known distance. We can then calculate its **average speed**, using the equation:

$$\text{average speed (metre per second, m/s)} = \frac{\text{distance travelled (metre, m)}}{\text{time taken (second, s)}}$$

Worked example

Ali's car took 20 s to travel 500 m.

His average speed $= \dfrac{500 \text{ m}}{20 \text{ s}} = 25$ m/s

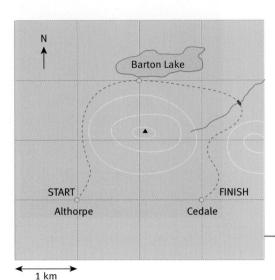

The distance from start to finish is 6 kilometres.

Worked example

Aisha runs a 100 metre race. The diagram shows her position at 1 second intervals during the race. She runs 100 metres in 12.5 seconds.

average speed $= \dfrac{100 \text{ m}}{12.5 \text{ s}} = 8$ m/s

But she didn't run at 8 m/s for the whole race, sometimes she ran more quickly, sometimes she ran more slowly, this is why we call this value the *average* speed.

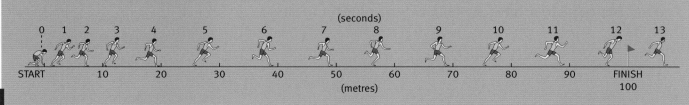

Instantaneous speed

The **instantaneous speed** of an object is the speed at which it is travelling at a particular instant. To estimate the instantaneous speed of an object, we measure its average speed over a very short time interval. The shorter we make this time interval, the less likely it is that the speed has changed much during it. On the other hand, if we make it too short, it is harder to measure the distance and the time accurately. As she crossed the finish line, Aisha ran approximately 10 metres in one second, so her *instantaneous* speed as she crossed the finish line was about 10 m/s.

Velocity and acceleration

People often use the words 'speed' and 'velocity' to mean the same thing. But the **velocity** of an object, also tells you the direction in which it is moving. So, for example, if a cyclist is pedalling at 8 metres per second along a road that runs to the west, her instantaneous speed is 8 m/s, but her instantaneous velocity is 8 m/s to the west.

In everyday language, if the speed of an object is increasing, we say that it is accelerating. Drivers are often interested in how quickly their car can accelerate. This might be stated as 0 to 60 miles per hour in 8 seconds; 0 to 60 mph is the change in speed and 8 s is the time the car took to change speed. In situations like this, we can use the equation:

$$\text{acceleration} = \frac{\text{change of speed}}{\text{time for the change to happen}}$$

Scientists use the change in velocity:

$$\text{acceleration (metre per second) per second, (m/s}^2\text{)} = \frac{\text{change of velocity (metre per second, m/s)}}{\text{time taken for the change (second, s)}}$$

Worked example

A car accelerates from 14 m/s to 30 m/s in 8 s. Calculate the acceleration.

change in speed = 30 m/s − 14 m/s = 16 m/s

$$\text{acceleration} = \frac{16 \text{ m/s}}{8 \text{ s}} = 2 \text{ m/s}^2$$

A car's speedometer measures its average speed over a very short time interval. Its reading is close to the instantaneous speed.

Summary box

- Average speed = $\frac{\text{distance travelled}}{\text{time taken}}$
- Instantaneous speed is the speed at a particular instant in time.
- Acceleration = $\frac{\text{change in speed}}{\text{time taken}}$

Questions

1. **a** How many seconds did Aisha take to run the first 20 m?
   b Calculate her average speed for the first 20 m.
2. A high-performance car can accelerate from 0–27 m/s in 6 seconds. Calculate the acceleration of the car in m/s².

F Picturing motion

Find out about

✓ how graphs can be used to summarize and analyse the motion of an object

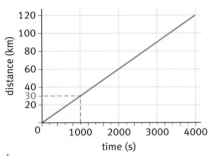

Distance–time graph for a car journey along the motorway. After 1000 s the car has travelled 30 km.

You can find out a lot about the motion of an object from a graph. A graph provides information about a moving object at any instant. From the shape and slope of the graph, we can work out how the object moved.

Distance–time graphs

The graph on the left is a distance–time graph for a car travelling along a motorway. Use the graph to find the distances the car has travelled after 1000 seconds, 2000 seconds, 3000 seconds, and 4000 seconds. The distance increases steadily so the speed is not changing. We say the speed of the car is constant. The constant **slope** of the distance–time graph is a straight line. This indicates a steady speed.

The graph below shows a cycle ride taken by Vijay. The graph has four sections, A, B, C, and D. In each section the slope of the graph is constant. This means that Vijay's speed is constant during that section of the ride.

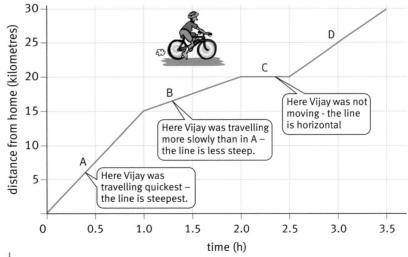

Vijay's cycle ride.

Questions

1. How far does Vijay travel during section D of his cycle ride?
2. How long does it take Vijay to cycle section D?
3. Which section does Vijay travel at the fastest speed?

Section A: In the first hour, Vijay travels 15 kilometres at a steady speed.

Section B: In the second hour he travels only 5 kilometres, because the route goes steeply uphill. The shallower slope indicates a lower steady speed.

Section C: In the third section (from 2.0 to 2.5 hours), his distance travelled does not change. He has stopped. This is what a horizontal section of a distance–time graph means.

This distance–time graph for Vijay's journey is not very realistic. In a real journey, the speed would change gradually rather than suddenly.

Speed–time graphs

The graph on the right is a speed–time graph for Vijay's trip on page 86. A steady speed is now shown by a straight horizontal line.

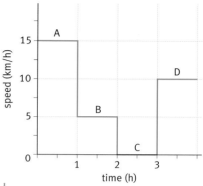

Speed–time graph of Vijay's cycle trip.

A more realistic speed–time graph would not have sudden changes of speed. There would be smoother, more gradual changes from one speed to another.

The second speed–time graph on the right is for a falling stone after 1 second its speed is 10 m/s. It has no horizontal sections, so the speed of the ball is changing all the time. The constant slope of the speed–time graph shows that the speed is changing at steadily. It has constant acceleration downwards.

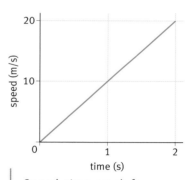

Speed–time graph for a falling stone.

> 4 Look at the speed–time graph of Vijay's cycle ride. What was his speed in
>   a section A?   c section C?
>   b section B?   d section D?
>
> 5
>
>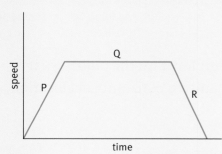
>
> Describe the motion shown by sections P, Q, and R of this graph.

G Forces and motion

Find out about

- force and acceleration
- momentum
- the link between change of momentum, force, and time

A force changes the motion of an object. The link is explored in these investigations.

Investigating force and motion

The diagram shows a small cart, with smooth well-oiled wheels, on a level table top.

The force of gravity on the hanging weight is constant, so the string exerts a steady force on the cart. The speed of the cart is measured at different time intervals after this pulling force is applied. You can see that the speed–time graph is a straight line. This tells us that the speed of the cart increases steadily with time.

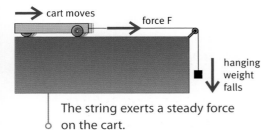

The string exerts a steady force on the cart.

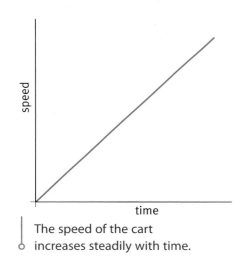

The speed of the cart increases steadily with time.

A constant force gives the cart a constant acceleration. As soon as the cart starts to move, its speed will keep on increasing as long as the force continues.

Freefall is another example of motion with constant acceleration. In this flash photograph of a falling ball the gaps between the positions of the ball keep increasing as it falls. It gets steadily faster. The speed–time graph is a straight line. The motion of a falling object is due to the gravitational force exerted on it by the Earth. This is a constant force on the object, so it causes its speed to increase steadily, a constant acceleration.

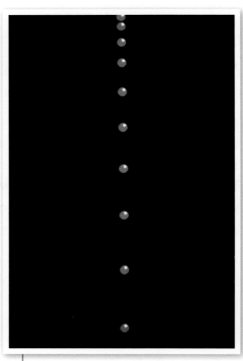

Freefall – a falling ball gets steadily faster as it falls.

Momentum

The **momentum** of a moving object is:

$$\text{momentum} = \text{mass} \times \text{velocity}$$
(kilogram metre per second, kg m/s) (kilogram, kg) (metre per second, m/s)

Summary box

- Force changes the motion of an object.
- A constant force causes a constant acceleration.

The faster an object is moving, the more momentum it has. A heavy object has more momentum than a lighter object moving at the same speed.

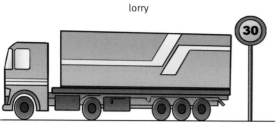

lorry
mass = 2800kg
velocity = 13 m/s
momentum = 36400 kg m/s

car
mass = 900kg
velocity = 13 m/s
momentum = 11700 kg m/s

motorbike
mass = 220kg
velocity = 31 m/s
momentum = 6820 kg m/s

Worked example
A whale of mass 5000 kg swimming at 2 m/s has momentum
= 5000 kg × 2 m/s = 10 000 kg m/s

The lorry in the diagram has more momentum than the car. The motorbike has less momentum than the car although it is travelling faster.

Investigating force and momentum
When two stationary objects spring apart, there is an interaction pair of forces. The forces are exactly the same size.

The diagrams show what happens when the mass of the objects is changed.

Question
1 What is the momentum of:
  a a skier of mass 50 kg moving at 5 m/s?
  b a netball of mass 0.5 kg moving at 3 m/s?

Summary box
✓ **Momentum = mass × velocity**

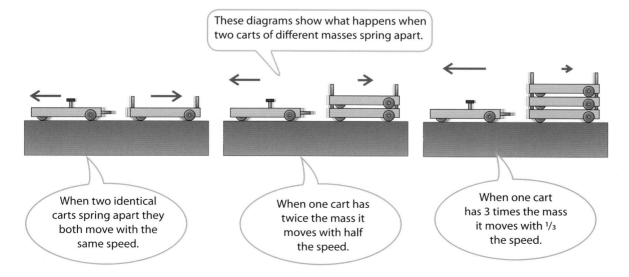

These diagrams show what happens when two carts of different masses spring apart.

When two identical carts spring apart they both move with the same speed.

When one cart has twice the mass it moves with half the speed.

When one cart has 3 times the mass it moves with ⅓ the speed.

Taking a free kick. The interaction between the footballer's foot and the ball causes a change of momentum.

Force and change of momentum

In an interaction like the three in the diagram on the previous page, the **change of momentum** is the same size for both objects. Two other things are also the same for both objects:

- the resultant force acting on the object (because the two forces in an interaction are always equal in size)
- the time for which it acts (the duration of the interaction, which has to be the same for both).

This is because there is a link between the force acting on an object, the time for which it acts, and the change of momentum it produces:

change of momentum of an object = resultant force acting on the object × time for which force acts
(kilogram metres per second, kg m/s)   (newtons, N)   (seconds, s)

Using the change of momentum equation

Questions

2 Which of the following will cause:
  i the largest change in momentum?
  ii the smallest change of momentum?

  a a force of 40 N acting for 3 s
  b a force of 200 N acting for 0.5 s
  c a force of 3 N acting for 50 s

Pushing a car

Ted and Tom push Karen's car. Together they exert a steady force of 800 N on the car. They push it for 10 seconds.

What is the change in momentum?

resultant force × time for which it acts = 800 N × 10 s = 8000 N s
So change of momentum of the car = 8000 kg m/s
The mass of the car is 1000 kg, so it must have reached a speed of 8 m/s.

Kicking a football

When a footballer takes a free kick, his foot exerts a force on the ball for a very short time.

This is the time that the foot and the ball are actually in contact. After that, the player's foot can no longer affect the motion of the ball. The kick has not given the ball some 'force' but it has given it some momentum.

Summary box

✓ Change in momentum = resultant force × time for which the force acts

Car safety

Cars today are much safer to travel in than cars ten or twenty years ago. Engineers use the results of crash tests like the one shown on the right to improve the designs. Newer, safer cars are still being designed changing.

Changing momentum safely

When a car is travelling at 100 km/h, the driver and passengers are also travelling at that speed. If the car comes to a very sudden stop, because of a collision, the occupants will continue travelling at 100 km/h until a force brings them to a sudden stop, like driver Hybrid 111 in the photograph.

The larger the force, the more serious the injuries. To reduce the damage we must reduce the force. We can do this by increasing the time it takes for the collision to make you stop.

Using the momentum equation for a collision:

change in momentum = stopping force × time for which the force acts.

To make this small…

…we must make this big.

A collision takes place over 0.5 s instead of 0.05 s. It takes 10 times as long, so the force is only one tenth of the size.

Crumple zones

Would you be safer in car **a** or car **b**?

Cars are fitted with front and rear crumple zones, with a rigid box in the middle. They are designed to crumple gradually in a collision. This makes it take longer for the rigid box to come to a stop. So the force on it is less.

> **Find out about**
> ✓ safety features in cars

Modern cars are designed to be safer.

This driver, Hybrid 111, has had lots of crashes. He has a steel skeleton and rubber skin and is packed with sensing equipment parts of the body. Each dummy costs more than £100,000.

1 0.00 s

2 0.05 s

3 0.10 s

4 0.15 s

5 0.20 s

6 0.25 s

7 0.30 s

Notice how the seatbelt stretches during the collision. So the change in momentum takes place over a longer time.

Seat belts and air bags

Some people think that seat belts work by stopping you moving in a crash. In fact, to work, a seat belt actually has to stretch. The film frames on the left show what happens to a driver wearing a seatbelt in a collision. As the seat belt stretches he moves forward, but more slowly than without a seatbelt. They make the change of momentum take longer. So the force is less. A crash helmet works in the same way. As the helmet deforms, it increases the time it takes for your head to stop, so the force is less.

Air bags cushion the impact in a collision. Again, they reduce your momentum over a longer time interval so that the force on you is less.

Could you save yourself?

Some people think they could survive a car accident without a seat belt, especially if they are travelling in the back seats. A car is travelling at 50 km/h (or roughly 14 m/s). Without a seat belt, the driver would hit the windscreen about 0.07 seconds after the impact. Back-seat passengers would hit the back of the front seats at roughly the same time. As your reaction time is typically about 0.14 seconds, this would all happen before you even have time to react.

Questions

1 When you jump down from a wall, it is a good idea to bend your knees as you land. Use the ideas on this page to explain why this reduces the risk of injury.

2 In railway stations, there are buffers at the end of the track. These are designed to stop the train if its brakes fail. The buffers are compressed when the train hits them. Explain how this would reduce the forces acting on the train and on the passengers.

Safety and risk

Car safety features, such as seat belts, are designed to reduce the **risk** of injury in a car accident. Travel can never be made completely safe. We cannot reduce the risk to zero. But many studies, in different countries, have shown that wearing seat belts greatly reduces the risk of serious injury. Many countries have regulations requiring drivers and passengers to use seat belts.

Questions

3 Wearing a seat belt is compulsory for front- and back-seat passengers in cars in Britain. A recent survey found that 94% of drivers and 95% of front-seat passengers wore seat belts. But only 69% of adult back-seat passengers wore a seat belt, though 96% of child back-seat passengers did.
  a Suggest a reason why a few drivers and front-seat passengers did not wear a seat belt.
  b Suggest a reason why 95% of front-seat passengers wore seat belts but only 69% of adult back-seat passengers.
  c Suggest a reason why 96% of children but only 69% of adults in back seats wore seat belts.

4 Look at the diagrams on the left.
  a After how long has the driver moved furthest forward?
  b The driver has a mass of 70 kg and was moving at 14 m/s before the collision. Calculate the change in his momentum during the collision.
  c Use your answer to part a to check that the force of the seat belt is about 4000 N.

Summary box
- The longer the stopping time in a collision, the smaller the force needed.

I Laws of motion

Find out about

- the laws (or rules) that apply to every example of motion
- how a resultant force is needed to change an object's motion

Steady speed

These diagrams show a player pushing a curling stone. He gives it momentum. What happens when he lets go depends on the friction of the surface.

On the 'perfect' ice there is no resultant force and the stone travels at a constant speed.

So motion at a steady speed does not need a force to maintain it. If the resultant force on an object is zero, its motion will remain unchanged. If it happens to be stationary, it will stay stationary. If it happens to be moving, it will carry on moving at the same speed in the same direction. Its momentum stays the same.

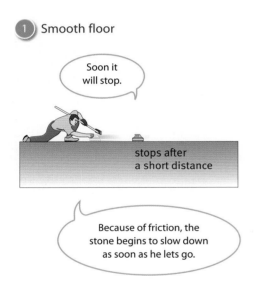

1 Smooth floor

"Soon it will stop."

stops after a short distance

Because of friction, the stone begins to slow down as soon as he lets go.

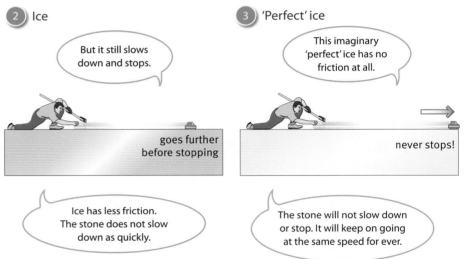

2 Ice

"But it still slows down and stops."

goes further before stopping

Ice has less friction. The stone does not slow down as quickly.

3 'Perfect' ice

"This imaginary 'perfect' ice has no friction at all."

never stops!

The stone will not slow down or stop. It will keep on going at the same speed for ever.

In the real world

In the real world there is always friction. So a driving force is needed to keep an object moving. The driving force has to balance the counter forces, like friction and air resistance, so that the resultant force is zero. When this occurs, the moving object does not slow down and stop. It keeps moving at whatever speed it had when the forces became equal.

Forces on a bicycle

When you pedal a bike there is a driving force. Air resistance and friction are the counter forces.

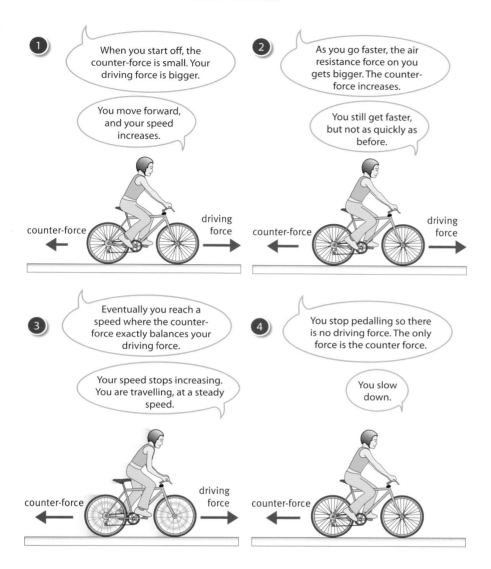

Summary box
- Newton worked out three laws of motion.
- If there is no resultant force acting on an object its momentum does not change.
- Change of momentum = resultant force × time for which it acts.
- The forces in an interaction pair are equal in size, but opposite in direction.

Slowing down

In image 4 the bike is moving forward but the force on it is in the opposite direction. Some people find it hard to believe that an object can move in one direction, whilst the resultant force acting on it is in the other direction. But this happens when an object is slowing down. Remember that a force does not cause motion; it causes a *change* of motion.

Questions

1. List three examples from everyday life of a situation where the resultant force on an object is zero.

2. List three examples from everyday life of a situation where there is a resultant force acting on an object.

J Work and energy

Find out about

- how to calculate the work done by a force
- the link between work done on an object and the energy transferred
- how to use energy ideas to predict the motion of objects

Transferring energy

When you push something and start it moving, your push transfers **energy** stored in your body (which gets less) to the energy of the moving object (which gets bigger). The energy of a moving object is called **kinetic energy**.

Work

In physics, the word 'work' has a special meaning.

When you push a trolley, you do more work if:
- it is hard to push
- you have to push it a long way.

So the amount of work depends on:
- the size of the force
- the distance the object moves in the direction of the force.

The equation for calculating the amount of work done by a force is:

work done by a force = force × distance moved in the direction of the force

(joule, J)   (newton, N)   (metre, m)

When you push the trolley, it speeds up. Its kinetic energy increases.

Worked example

Calculate how much work is needed to push a car 50 m along a road.

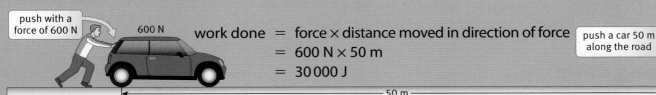

work done = force × distance moved in direction of force
= 600 N × 50 m
= 30 000 J

Summary box

- Energy is needed to do work.
- Work done = force × distance moved by force.

Question

1 It takes a force of 1200 N to push a large car along the road. How much work would you have to do to push it 40 m?

Work and energy

amount of energy = amount of work done
   transferred   by the force

Energy and work are both measured in joules (J). A force of 1 newton applied over a distance of 1 metre does 1 J of work, and transfers 1 joule of energy.

Not working!

If you push an object and cannot get it to move, you feel that you are doing hard work! From the physics point of view, you are not doing work because the force you are applying is not moving anything. Holding a heavy object also feels like hard work, but the force is not making it move, so you are not doing any work.

Lifting things: gravitational potential energy

You do work when you lift luggage into the car boot. When you transfer energy from your body's store of chemical energy, the **gravitational potential energy** of the luggage increases. The increase is equal to the amount of work you have done.

In general, when anything is lifted up, you can calculate the change in gravitational potential energy from the equation

gravitational potential energy = weight × vertical height difference
        (joule, J)   (newton, N)   (metre, m)

Notice that it is only the vertical height difference that matters. If you slide a suitcase up a ramp, the gain in gravitational potential energy is the same as if you lift it vertically.

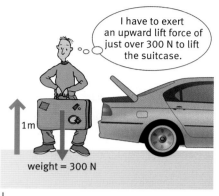

Doing work by lifting: increasing gravitational potential energy.

Worked example

A suitcase weighs 300 N. To lift it up needs an upward force of 300 N. To lift it 1 metre into the boot of the car, then

work done = force × distance moved in the direction of the force

= 300 N × 1 m

= 300 J

The suitcase now has 300 J more gravitational potential energy.

Question

2 Jade's weight is roughly 400 N. How much work does she do when she goes upstairs – a vertical height gain of 2.5 m?

Summary box

- Work done = energy transferred.
- Work and energy are measured in joules.

a

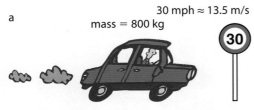

30 mph ≈ 13.5 m/s
mass = 800 kg

kinetic energy = $\frac{1}{2}$ × 800 kg × (13.5 m/s)^2
= 72 900 J

b

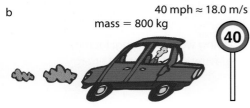

40 mph ≈ 18.0 m/s
mass = 800 kg

kinetic energy = $\frac{1}{2}$ × 800 kg × (18 m/s)^2
= 129 600 J

A car travelling at 40 mph (**b**) has nearly twice as much kinetic energy as the same car at 30 mph (**a**). **b** uses a lot more fuel to get to this speed.

Questions

3. A ten-pin bowling ball has a mass of 4 kg. It is moving at 8 m/s. How much kinetic energy does it have?

4. Which of the following has more kinetic energy?
   a. a car of mass 500 kg travelling at 20 m/s
   b. a car of mass 1000 kg travelling at 10 m/s

Summary box

- Kinetic energy = $\frac{1}{2}mv^2$
- Change in kinetic energy = work done by a force.

Calculating kinetic energy

The equation for calculating the **kinetic energy** of a moving object is:

kinetic energy = $\frac{1}{2}$ × mass × (velocity)^2
(joule, J)   (kilogram, kg) (metre per second, m/s)^2

The amount of kinetic energy depends on the mass and the velocity squared. Small changes in velocity can mean big changes in kinetic energy.

Making things speed up: changing their kinetic energy

When Anna pushes a supermarket trolley along a level floor, she is doing work. The trolley's speed increases as long as she keeps pushing. She is transferring energy to the trolley, where it is stored as kinetic energy. If the trolley has no friction, the amount of work she does pushing it is equal to the change in the trolley's kinetic energy.

Worked example

Calculate the change in kinetic energy of a trolley pushed with a force of 6 N over a distance of 5 m (assume there are no frictional forces acting).

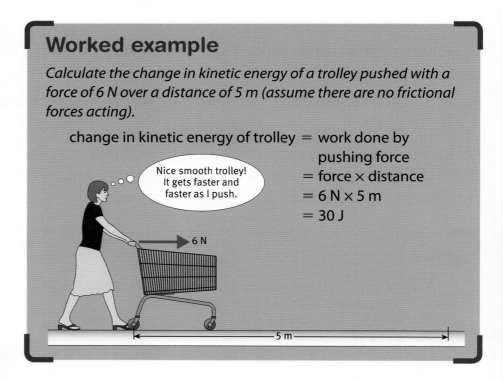

change in kinetic energy of trolley = work done by pushing force
= force × distance
= 6 N × 5 m
= 30 J

A real trolley always has some friction, so its change in kinetic energy will be less than this. Some work is wasted as heat.

Worked example

Calculate the gain in kinetic energy of the roller coaster shown below at the bottom of the ride (assuming there are no friction forces).

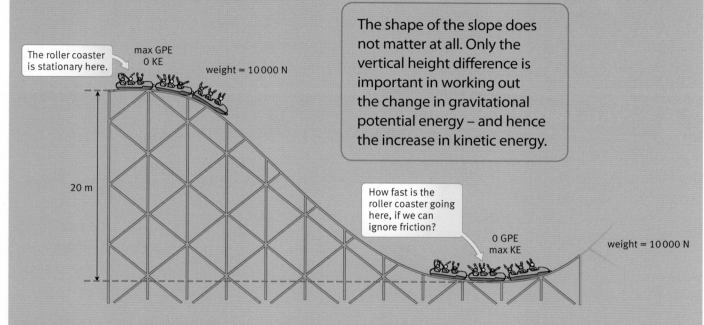

If friction is small enough to ignore then the loss of gravitational potential energy is equal to the kinetic energy it gains:

loss of gravitational potential energy = weight × vertical height change
= 10 000 N × 20 m
= 200 000 J

loss of gravitational potential energy = gain in kinetic energy
So gain in kinetic energy = 200 000 J

Conservation of energy

When something falls through a certain height difference, or slides down a smooth ramp, the amount of gravitational potential energy (GPE) stored in the system gets less, and the kinetic energy (KE) of the moving object gets bigger, if friction is small enough to be ignored.

This doesn't depend on the shape of the path the object follows, as long as it is smooth enough to allow us to ignore friction. A fairground roller coaster is a good example of these ideas in action.

Even if the track has a complicated shape, we can use the principle of **conservation of energy**.

Summary box
- If friction is small enough to be ignored, for an object moving upwards, gain in GPE = Loss in KE.
- For an object moving downwards, Loss in GPE = gain in KE.

Science Explanations

Forces and motion form the basis of our understanding of how the world works. Every example of motion we observe can be explained by a few simple rules that are so exact and precise that they can be used to predict the motion of an object very accurately.

You should know:

- about the pair of forces that always arise when two objects interact
- that vehicles move by pushing back on something and this interaction causes a forward force to act on them
- about the friction between two objects that are sliding past each other
- that the resultant force on an object is the sum of all the individual forces acting on it, taking their directions into account
- about the reaction force on an object that arises because it pushes down on a surface
- how to calculate the average speed of a moving object
- about instantaneous speed and how this is different to the average speed
- what is meant by distance, speed, velocity, and acceleration
- how to draw and interpret distance–time and speed–time graphs
- how to calculate the acceleration of an object
- that when a resultant force acts on an object it causes a change in momentum
- how to calculate momentum and the force that causes a change in momentum
- that many vehicle safety features increase the time for an event (such as a collision), so that the average force is less, for the same change of momentum
- that if there is no resultant force on an object, its momentum does not change – it either remains stationary or keeps moving at a steady speed in a straight line
- about the work done when a force moves an object and how to calculate the work done, which is equal to the energy transferred
- that when work is done on an object, energy is transferred to the object and when work is done by an object, energy is transferred from the object to something else
- about gravitational potential energy and how to calculate the change in gravitational potential energy as an object is raised or lowered
- about kinetic energy, how to calculate it, and that doing work on an object can increase its kinetic energy by making it move faster
- that when a object falls, if friction and air resistance can be ignored, the decrease in gravitational potential energy is equal to the increase in kinetic energy.

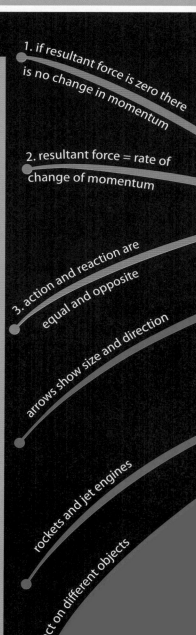

1. if resultant force is zero there is no change in momentum

2. resultant force = rate of change of momentum

3. action and reaction are equal and opposite

arrows show size and direction

rockets and jet engines

act on different objects

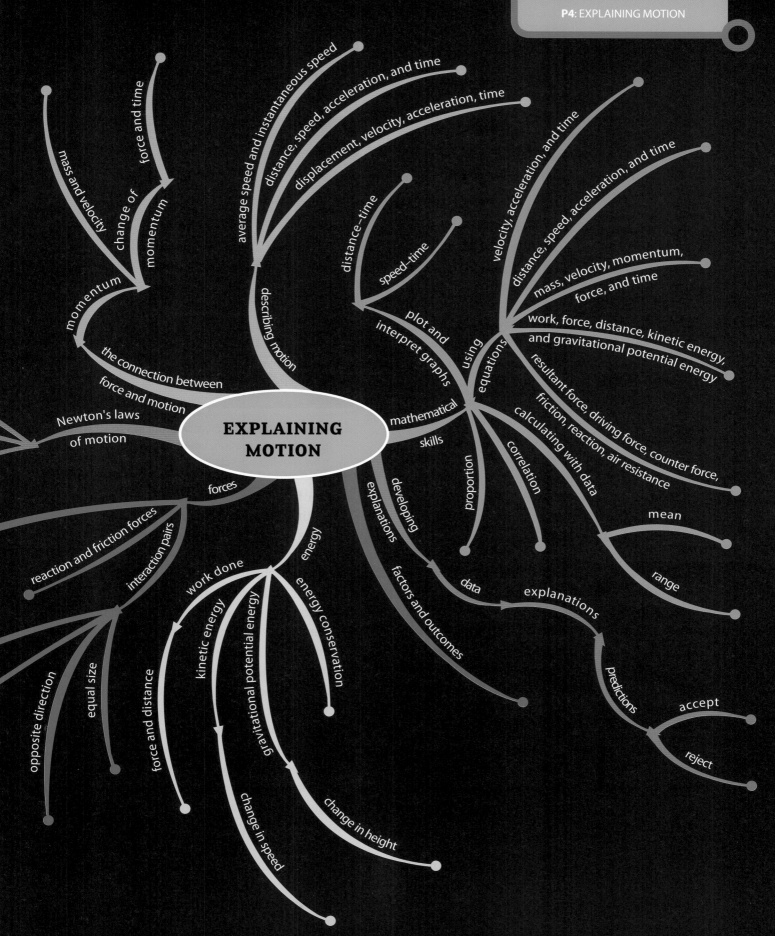

Ideas about Science

In addition to understanding forces and motion, you need to understand how scientific explanations are developed. The list below links these ideas about science with some examples from the module.

In developing scientific explanations you should be able to:

- identify statements that are data and statements that are explanations. For example, the statement 'the acceleration of a falling object has a constant value' is data; the statement 'when a constant force acts on an object it causes a constant acceleration' is an explanation.
- recognise that an explanation may be incorrect, even if the data is correct. For example, some measurements show that pushing with a constant force on a toy car causes it to travel at a constant speed, but the explanation 'a constant force causes a car to travel at a constant speed' is incorrect. The pushing force is balancing the friction force – there is no resultant force on the car.
- identify where creative thinking is involved in the development of an explanation. For example, the idea of an interaction pair of forces means that when you push on a wall the wall pushes back – this is not an obvious idea, but Isaac Newton suggested this as his third law of motion: action and reaction are equal and opposite.
- recognise data or observations that are accounted for, or conflict with, an explanation. For example, if data shows that the force on an object is proportional to its acceleration, this agrees with Newton's second law: the force on an object is equal to its rate of change of momentum.
- give good reasons for accepting or rejecting a scientific explanation. For example, Aristotle believed that heavier objects always fall

Isaac Newton used creative thinking to write his three laws of motion, which explain and predict how moving objects behave.

faster than light objects. A good reason for rejecting this would be a slow-motion film of objects falling in a vacuum – they fall at the same speed.

- decide which of two scientific explanations is better. Galileo said that light and heavy objects fall at the same speed. This matches observations better, when there is no air resistance.
- understand that when a prediction agrees with an observation this increases confidence in the explanation on which the prediction is based, but does not prove it is correct. For example, Newton's laws of motion have correctly predicted the behaviour of many moving objects.
- understand that when a prediction disagrees with an observation this indicates that one or the other is wrong and decreases confidence in the explanation on which the prediction is based. For example, observations do not support Aristotle's ideas about motion.

Review Questions

P4: EXPLAINING MOTION

1 Think about the following situations:
   a Amjad on his skateboard, throwing a heavy ball to his friend (main objects to consider: Amjad, the skateboard, and the ball)
   b a furniture remover trying to pull a piano across the floor, but it will not move (main objects to consider: the furniture remover, the piano, and the floor)
   c a hanging basket of flowers outside a café (main objects to consider: the basket and the chain it is hanging from).

For each situation:
   a Sketch a diagram (looking at it from the side).
   b Sketch separate diagrams of the main objects in the situation (these are listed for each).
   c On these separate diagrams, draw arrows to show the forces acting on that object. Use the length of the arrow to show how big each force is.
   d Write a label beside each arrow to show what the force is.

2 A tin of beans on a kitchen shelf is not falling, even though gravity is still acting on it. The shelf exerts an upward force, which balances the force of gravity. Explain in a short paragraph how it is possible for a shelf to exert a force. Sketch a diagram to help your explanation.

3 a The winner of a 50-m swimming event completes the distance in 80 s. What is his average speed?
   b A car travels 280 miles from London to Newcastle in 5 hours.
      i What is the average speed of the car?
      ii Explain how you know that the car must have travelled faster than this at some time.

4 a A bus leaves a bus stop and reaches a speed of 15 m/s in 10 s. Calculate its acceleration.
   b At the 2008 Olympic games Usain Bolt accelerated to a speed of 12 m/s after 6 seconds. What was his average acceleration?

5 The table shows the distance run and time taken for a sprinter in training.

time in s	0	1	2	3	4	5	6	7	8	9	10	11
distance in m	0	3	9	18	27	36	45	50	55	60	65	70

   a Plot the data on a graph.
   b Label your graph to show where the sprinter is running fastest.
   c Label the graph to show where he is running slowest.

6 What is the momentum of:
   a a hockey ball of mass 0.4 kg moving at 5 m/s?
   b a jogger of mass 55 kg, running at 4 m/s?
   c a van of mass 10 000 kg, travelling at 15 m/s?
   d a car ferry of mass 20 000 000 kg, moving at 0.5 m/s

7 A weightlifter raises a bar of weight 500 N until it is above his head – a total height gain of 2.2 m.
   a How much gravitational potential energy has the bar gained?
   b He drops it from 2.2 m. How much kinetic energy will it have when it hits the floor?

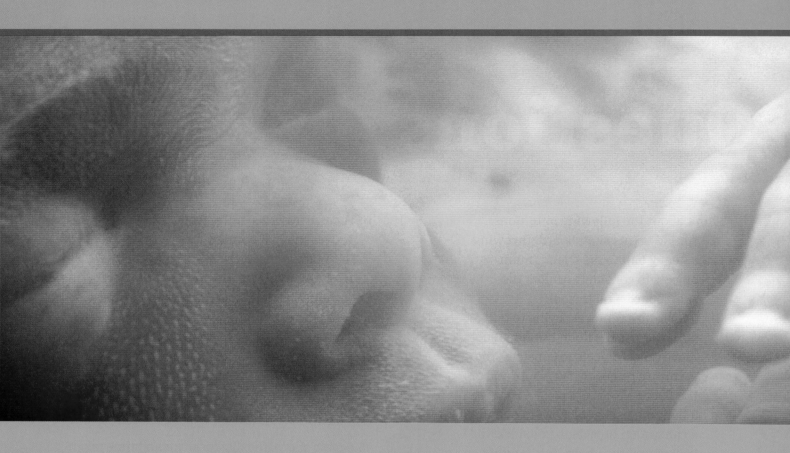

B5 Growth and development

Why study growth and development?

How does a human embryo develop? What makes cells with the same genes develop differently? Exploring questions like these is part of the fast-moving world of modern biology.

What you already know

- Genes affect the way living things develop.
- Clones, such as plants grown from cuttings, are more similar to each other than organisms with a combination of their parents' genes.
- Living things are made up of cells, and these divide and develop into the whole organism.

Find out about

- the structure of DNA, and how it controls the proteins a cell makes
- how cells divide to make sex cells and to make new body cells
- how cells become specialised
- the differences between plant and animal growth.

The Science

Sex cells carry DNA – the genetic information to make a new individual. Cells become specialised because of the different proteins they make. DNA's unique structure determines the proteins a cell makes.

Ideas about Science

Data and explanations of data are different. Creative thinking is needed to develop explanations from data.

A | Growing and changing

Find out about
- different cells, tissues, and organs
- growing up

You began life as a single cell. By the time you were born you were made of millions of cells. You probably weighed about 3 or 4 kilogrammes. Now you probably weigh over 50 kilogrammes. Not only have you grown, but you have changed in many ways. In other words, you have developed.

Since you were born, your **development** has been gradual. In some plants and animals, development involves big changes. For example, the young and the adults in these photographs look very different.

Questions

1 Match the young and the adults in pictures A to F.

2 Which animal in the pictures has a life cycle most like a fly?

The life cycle of the fly

Until 1668 people thought that wriggling maggots just appeared in old meat. The scientist Francesco Redi did experiments to show that the maggots came from flies. Maggots only appeared where flies could land on the meat.

- Flies lay eggs.
- Eggs hatch into maggots.
- Maggots change into pupae. Inside the pupa, the tissues reorganise into a fly.

Cells are the building blocks of living things

The plants and animals in the pictures opposite are made of lots of cells – they are multicellular. Your body has more than 300 different kinds of cell. Each kind of cell is **specialised** to do a particular job.

Cells make tissues, tissues make organs

All newly formed human cells look much the same. Then they develop into groups of specialised cells called **tissues**.

Plant cells are different from animal cells, but they are specialised too. Plant cells have cell walls outside the cell membrane, and some have spaces called vacuoles.

As an animal embryo or plant grows, groups of tissues arrange themselves into **organs**, for example, the heart and brain in humans, and roots, leaves, and flowers in plants.

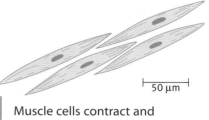

Muscle cells contract and relax to cause movement.

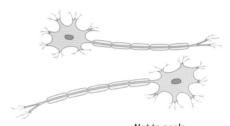

Not to scale
Nerve cells carry nerve impulses.

Different tissues in a leaf – a plant organ.

- waxy cuticle
- epidermis
- palisade layer
- tissue for photosynthesis
- **xylem** tissue for transport of water and minerals
- **phloem** tissue for sugar transport
- spongy layer

Summary box
- ✓ All living things are made of cells.
- ✓ Specialised cells join up to make tissues like muscles and nerve tissue.
- ✓ Organs are made of groups of different tissues.

Questions

3 What does 'specialised' mean?

4 Read about Francesco Redi's experiment on the previous page. Sum up what you think he did, in a flowchart. Write down what you think he saw. What conclusions did he make?

5 Explain the difference between a tissue and an organ.

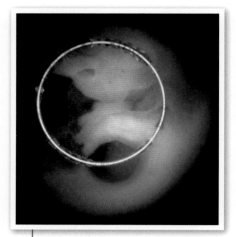

A human fetus at eight weeks. This photograph was taken from inside the mother's uterus. The fetus is about 2.5 cm long.

From single cell to adult

All the cells in your body come from just one original cell – a fertilised egg cell or **zygote**.

So the zygote must contain instructions for making all the different types of cells in your body, for example, muscle cells, bone cells, and blood cells. It also has the information to make sure that each type of cell develops in the right place and at the right time. This information is in your DNA – the chemical that your genes are made of.

In humans:

sperm + egg cell →(fertilisation)→ zygote (fertilised egg cell) →(division of cells by mitosis)→ embryo

The growing baby

During the first week of growth, the zygote divides to form a ball of about 100 cells. The nucleus of each cell contains an exact copy of the original DNA. As the embryo grows, some of the new cells become specialised and form tissues. After about two months, the main organs have formed. The developing baby is now called a **fetus**.

When the embryo is a ball of eight cells or fewer, it occasionally splits into two. A separate embryo develops from each section. When this happens, identical twins are produced. They are **clones** of each other. This shows that there are cells in the early embryo that are identical and unspecialised. These cells can develop into complete individuals. They are called **embryonic stem cells**.

Summary box

- ✓ When an egg cell is fertilised by a sperm cell, a zygote is formed.
- ✓ A zygote divides to form an embryo.
- ✓ Up to the eight-cell stage an embryo is made of unspecialised embryonic stem cells.
- ✓ After about two months a developing human body has organs and is called a fetus.
- ✓ Plants have meristem cells that keep dividing, allowing the plant to grow.

Questions

6 What is a zygote?

7 When does a human embryo become a fetus?

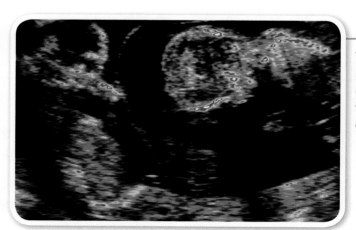

This ultrasound scan shows that twins are expected. Both of the babies' heads can be seen.

Growth patterns

For living things to grow bigger, some of their cells must divide to make new cells. You will probably stop growing taller by the time you are about 18–20 years old.

Flowering plants continue to grow throughout their lives.
- Their stems grow taller.
- Their roots grow longer.
- Their stems get wider.

Plants increase in length by making new cells at the tips of both shoots and roots. They also have rings of dividing cells in their stems and roots to increase their thickness. These dividing cells are called **meristem cells**.

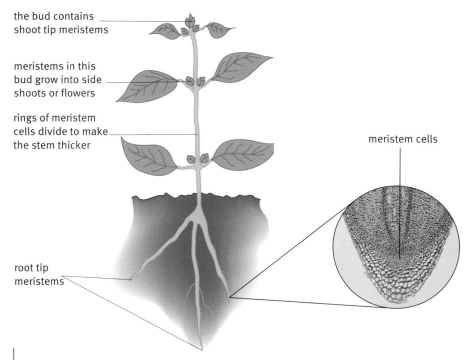

This giant sequoia tree is over 2000 years old, 83 m tall, and 26 m in girth (circumference). It is known as 'General Sherman' and is officially the largest giant sequoia tree and the largest living thing on Earth.

Meristem cells divide to make stems and roots longer and make the stem thicker. On the right is a root tip meristem (photographed through a microscope).

Questions

8 Why is it important that all living things have cells that can divide?

9 Name the type of cell in a plant that can divide to make new tissues.

B | Growing plants

Find out about

- why plants are so good at repairing damage

Cells in your body divide when you are growing. If you cut yourself, cells divide to repair your body, but only for small repairs. Many plants and some animals can replace whole organs.

Why can plants grow back?

Plants keep some meristem cells all through their lives. Plant meristem cells are **unspecialised**. These spare back-up cells can divide to make any kind of cell the plant needs. So plants can regrow whole organs, such as leaves, if they are damaged.

Stem cells can make new specialised cells

Animals also have spare back-up cells called **stem cells**. These cells divide, grow, and develop into specialised cells the body needs.

1 — normal front leg

2 — leg bitten off by a predator

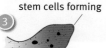
3 — stem cells forming / leg after 3 weeks

4 — leg after 4 weeks

5 — leg after 6 weeks

6 — leg after 10 weeks

If a newt's leg is bitten off, it can grow a new one. Most animals can only make small repairs to their body.

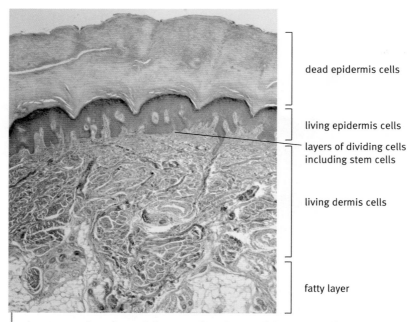

A cross-section through human skin. Some of the stem cells continue to grow and divide. Others replace skin cells at a wound or those that wear off at the surface (×36).

Labels: dead epidermis cells; living epidermis cells; layers of dividing cells including stem cells; living dermis cells; fatty layer

Newts' stem cells stay unspecialised throughout their lives. So newts can grow new legs if they need to – or even an eye.

The stem cells in adult humans are not as useful, because they have already started to specialise. For example, the stem cells in your skin can only develop into skin cells.

Using meristems to make more plants

In module B1, you saw that gardeners use meristems when they grow new plants by taking **cuttings**. Cuttings are just shoots or leaves cut from a plant. In the right conditions they develop roots and grow into new plants.

Cuttings grow better when you dip the cut ends in **rooting powder** before you plant them. Rooting powder contains plant hormones that cause some of the shoot meristem cells to develop into roots.

By taking cuttings, gardeners can produce lots of new plants quickly and cheaply. All the cuttings taken from one plant have identical DNA. As they are genetically identical, they are called **clones**. So taking cuttings is a good way of reproducing a plant with exactly the features that you want. When flowering plants produce seeds, they are reproducing sexually. So new plants grown from seeds vary. They are not identical.

Rooting powder encourages new roots to grow.

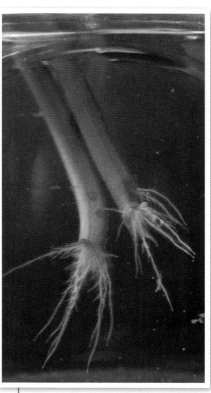
This willow cutting needed only water to grow new roots.

Summary box
- Plants have unspecialised meristems that can grow into new specialised cells.
- Animal stem cells can make new cells.
- Plant meristems form new roots or shoots when a gardener takes a cutting.
- Rooting powder contains hormones that encourage root growth.
- Cuttings are genetically identical to each other.

Questions
1. Name two parts of your body where you have stem cells.
2. For each of these types of cell, say whether they are fully unspecialised or not:
   a. meristem cells
   b. embryonic human stem cells
   c. adult human stem cells.
3. Explain why a newt can re-grow a leg but a human cannot.
4. Give two reasons for growing plants from cuttings.

C Phototropism

Find out about

- why plants grow towards light

Conditions in the environment affect how well a plant grows and develops. Plants rooted in soil cannot move from place to place – not even the 'walking palm' tree in the picture below.

This houseplant has grown towards the window to increase the amount of light falling on its leaves.

The walking palm tree, *Socratea durissima*, in Costa Rica. New roots grow towards a sunny patch and pull the stem and leaves towards the light. Hormones control the direction of root growth. Older roots in the shade die.

You may have noticed that plants on windowsills seem to bend towards the light. They are not moving, but growing. When plants grow towards light, this is called **phototropism**. Phototropism increases a plant's chances of survival.

Questions

1. Write a definition for phototropism.
2. How is the walking palm tree increasing its chance of survival?
3. Explain why a plant benefits from growing towards the light.

Darwin's phototropism experiments

Charles Darwin experimented with phototropism. He showed that the young shoots of grasses:
- normally grew towards light
- remained straight when he covered their tips.

Summary box
- Plant stems grow towards the light - this is called phototropism.

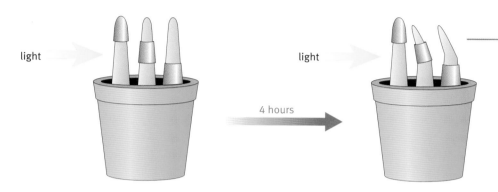

Foil covers different parts of the barley shoots.

In the experiment, shown in the picture above, covering the lower parts of the shoot did not stop it growing towards the light. This shows that only the tip is sensitive to light. The shoot bends below the tip – where cells are no longer dividing but are increasing in length.

Darwin did not know how bending towards light happened. Now scientists have found that higher concentrations of plant hormones cause shoot cells to grow faster on the shady side.

Questions

4 Look at the diagram showing phototropism experiments in barley. Suggest where:
  a the shoot detects light
  b the cells are growing very quickly to cause bending.

5 Apart from light, what other factors do you think will affect how well a plant grows and develops?

D A look inside the nucleus

Find out about

- where genes are kept inside your cells

All cells start their lives with a nucleus. You can see a cell's nucleus using a microscope. A few special cells lose their nuclei when they finish growing. Human red blood cells are one example. Their job is to carry oxygen. They have more space to carry oxygen without a nucleus.

Summary box

- Chromosomes are found inside the nucleus.
- Chromosomes are made of DNA, which can copy itself.
- DNA is arranged into thousands of genes.
- Each gene has the information to make one protein.

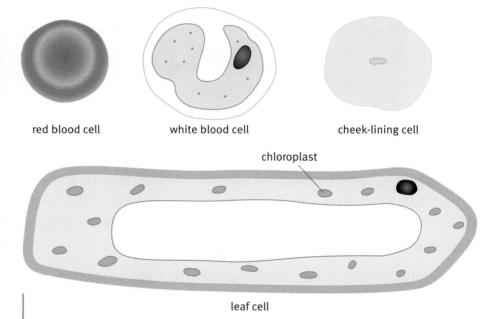

Cells vary in size and shape.

Organism	Estimated gene number	Chromosome number
human	~30 000	46
mouse	~30 000	40
fruit fly	13 600	8
Arabidopsis thaliana (plant)	25 500	5
roundworm	19 100	6
yeast	6 300	16
Escherichia coli (bacterium)	3 200	1

Chromosomes contain genes

A **chromosome** is a long molecule of DNA wound around a protein framework. You have about a metre of DNA in each of your nuclei. This is made up of about 30 000 genes. Each gene codes for a protein.

Different species have different numbers of chromosomes and different numbers of genes (see the table on the left).

You have 23 pairs of chromosomes in your nuclei. You got one set of 23 from your mother's egg cell nucleus and the other set from the nucleus of your father's sperm cell.

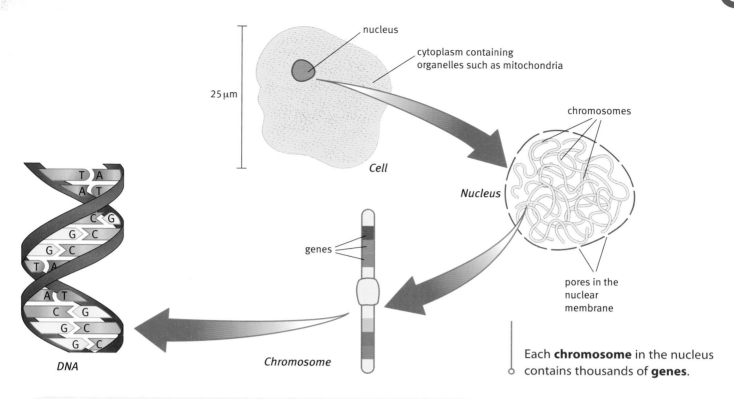

Each **chromosome** in the nucleus contains thousands of **genes**.

These people are more alike than it appears. 99.9 % of their genes are the same.

DNA is a special molecule

DNA has a special structure that allows it to:
- make exact copies of itself
- provide instructions for making proteins.

Questions

1. Name two ways in which your red blood cells are different from the other cells in your body.
2. How many different types of protein are there likely to be in yeast?
3. Which organism has about half as many genes as yeast?
4. Describe DNA's special properties.

E | Making new cells

Find out about

- how your cells divide for growth and repair your body

Imagine a space probe bringing back objects like this from Mars. Scientists would need to find out whether they were alive or not. It might be a living organism able to colonise Earth!

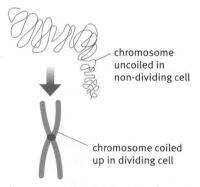

You can see chromosomes clearly only in dividing cells. The rest of the time the long DNA molecule is unwound.

It is not always easy to tell whether something is living or not.

But life cannot exist without the growth, repair, and reproduction of cells.

Cells grow and divide

When new body cells are made, they contain the same number of chromosomes as each other and the parent cell. They also contain the same tiny cell parts, such as chloroplasts and **mitochondria**. These are called **organelles**. So, before a cell divides, it must grow and make copies of:

- its organelles
- its chromosomes.

Only then does the cell divide. This type of cell division is called **mitosis**.

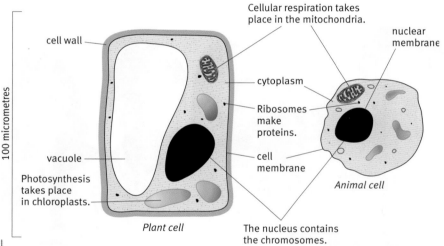

Cell organelles, such as chloroplasts and mitochondria, are copied before a cell divides.

Copying chromosomes

You can see the chromosomes in dividing cells by using a light microscope. The DNA is too spread out in other cells to be visible. After the chromosomes are copied, the DNA strands become shorter and fatter.

Body cells divide by mitosis

Most cells in your body divide by mitosis. Once the chromosomes have been copied the cell can split in two.

First, a complete set of chromosomes goes to each end of the dividing cell. Two new nuclei are made. A complete set of organelles also goes to each end. Then the cytoplasm divides. Two genetically identical cells are formed. In plant cells, a new cell wall forms too.

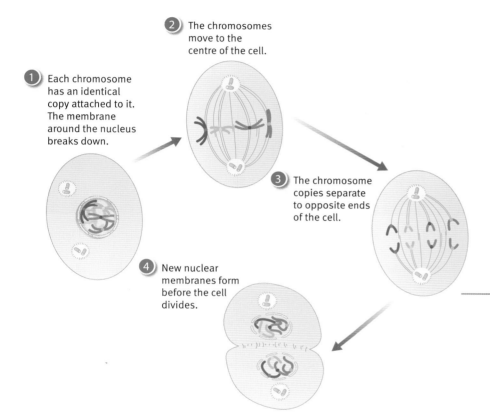

1. Each chromosome has an identical copy attached to it. The membrane around the nucleus breaks down.
2. The chromosomes move to the centre of the cell.
3. The chromosome copies separate to opposite ends of the cell.
4. New nuclear membranes form before the cell divides.

Summary box
- Body cells divide by mitosis to make two new genetically identical cells.
- The chromosomes need to be copied before the cell divides.
- The copied chromosomes are separated in an organised way.
- Some plants and simple creatures can reproduce on their own asexually using mitosis.

The pictures show what happens when a cell divides by mitosis. They show only four chromosomes.

Mitosis and asexual reproduction

Some plants and animals reproduce on their own asexually. They use mitosis to produce cells for a new individual.

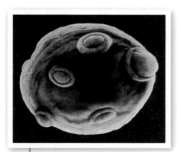

In yeast, new cells grow as buds from the parent.

Daffodil bulbs divide to form new identical ones.

Each of the offspring produced in asexual reproduction is genetically identical to the parent. It is called a clone of its parent.

Questions

1. a How many cells are made by mitosis?
   b What are these new cells like compared with their parent cell?
2. Before a cell divides, it grows. What two steps happen during cell growth?
3. What two main steps happen during cell division by mitosis?

F Sexual reproduction

Find out about

✓ cell division to make gametes

Most plants and animals reproduce sexually. Males and females make sex cells or **gametes**, which join up at fertilisation. The fertilised egg, or zygote, develops into the new life.

Male and female gametes are different

Only the male peacock has magnificent tail feathers.

The berries show that the holly on the right is female. You cannot be sure about the one on the left.

A snail has both male and female sex organs.

The only way to be sure about the sex of a living thing is to look at its gametes. Male gametes are usually made in very large numbers. They move to the female gamete by swimming or being carried by the wind or an insect. Females have large gametes that stay in one place.

Human males produce sperm in their testes. Females produce egg cells in their ovaries.

250 micrometres

Sperm develop in tubules in the testes.

600 micrometres

Pollen contains the male gametes of a flowering plant.

Gametes have half the chromosomes of a normal cell

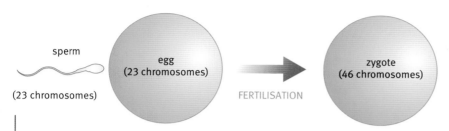

Meiosis halves the number of chromosomes in the sex cells. Fertilisation restores the number in the zygote.

Human body cells have 23 pairs of chromosomes – 46 in total. Gametes have only 23 single chromosomes. When a sperm cell fertilises an egg cell, their nuclei join up. The fertilised egg cell (zygote) gets the correct number of chromosomes: 23 pairs – 46 in total. Half your chromosomes come from your mother and half come from your father. Gametes are made by a special kind of division called **meiosis**.

Gametes are made by meiosis

Meiosis only happens in sex cells. In humans, meiosis makes gametes that:
- have 23 single chromosomes (one from each pair)
- are all different – no two gametes have exactly the same genetic information.

Offspring from sexual reproduction are different from each other and from their parents. We say that they show **genetic variation**.

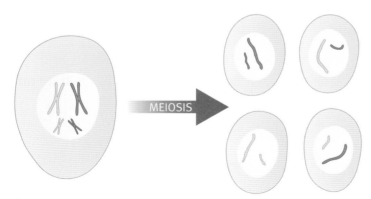

The chromosomes in this diagram have been copied. The parent cell divides twice, producing four cells. They have half the number of chromosomes as the parent cell. No two gametes have exactly the same genetic information.

B5: GROWTH AND DEVELOPMENT

Summary box
- Sexual reproduction gives useful genetic variation.
- Male and female gametes are different.
- Gametes are made by a special type of cell division called meiosis.
- Meiosis halves the chromosome number and introduces variation.
- Fertilisation restores the chromosome number.

Questions

1. Look again at the photos of holly at the top of the opposite page. You can be sure that the holly on the right is female. Why can you not be sure about the sex of the holly on the left?

2. Why is it important that gametes have only one set of chromosomes?

3. Why are male gametes made in such large numbers?

F: SEXUAL REPRODUCTION

G | The mystery of inheritance

Find out about

- the structure of DNA
- how DNA is copied for cell division

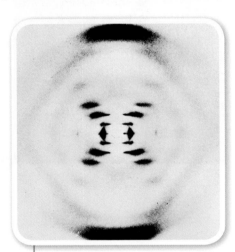

X-ray diffraction pictures of DNA, like this, show that DNA has a repeating pattern.

Watson (left) and Crick reveal their model of DNA.

In 1865 Gregor Mendel published his work on pea plants. Mendel's data showed how features could be passed on from parents to their young. The explanation for how information is passed on did not simply emerge from Mendel's data. It took the work and creative thinking of many scientists to explain how it happens.

Working out the structure of DNA

1859	A chemical was extracted from nuclei and named 'nuclein'.
1944	'Nuclein' was recognised as genetic material.
Late 1940s	Erwin Chargaff discovered a pattern in the number of bases in DNA.
1951	Linus Pauling and Robert Corey showed that proteins have a helix structure.
1952–3	Rosalind Franklin and Maurice Wilkins produced X-ray diffraction pictures of DNA. They showed that the molecule had a regular, repeating structure.
1953	Francis Crick and James Watson work out the spiral double-helix structure of DNA.

Some of the discoveries that led up to the discovery of DNA structure. These discoveries provided explanations that accounted for Mendel's data. Predictions based on the explanations were then tested by further experiments.

Base pairing is the key to how DNA works

There are four special molecules called 'bases' in DNA: adenine (A), thymine (T), guanine (G), and cytosine (C).

The amount of A is always the same as the amount of T, and the amount of G is the same as the amount of C. This is true no matter what organism the DNA comes from.

Crick and Watson concluded from this evidence that:
- A always pairs with T.
- G always pairs with C.

This is **base pairing**.

The DNA double helix

The bases fit together like the rungs of a ladder. The other DNA chemicals make long chains like the sides of a ladder. The whole molecule is twisted into a spiral shape called a double helix.

How DNA can be copied

Base pairing means that it is possible to make exact copies of DNA:
- Weak bonds between the bases split, unzipping the DNA from one end to form two strands.
- Immediately, new strands start to form from free bases in the cell.
- As A always pairs with T, and G always pairs with C, the two new chains are identical to the original.

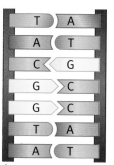

The rungs of the ladder are the pairs of bases held together by weak chemical bonds.

There are ten pairs of bases for each twist in the helix.

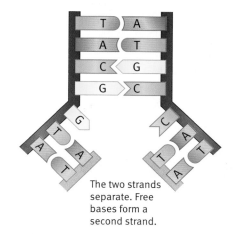

The two strands separate. Free bases form a second strand.

Each DNA molecule is made of half old DNA (black) and half new DNA (grey).

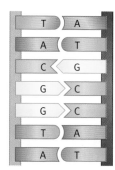

Each half of the split DNA molecule is complete, making two identical DNA molecules.

The structure of DNA (simplified).

Questions

1. The shape of a DNA molecule is sometimes described as a twisted ladder.
   a What makes the sides of the ladder?
   b What part of the ladder are the bases?

2. The bases in a DNA molecule always pair up the same way. Which base pairs with:
   a A?   c G?
   b C?   d T?

3. Describe what happens when a DNA molecule is copied.

H Making proteins

Find out about

- how DNA controls which protein a cell makes

Proteins are long chains of amino acids

Proteins do lots of different jobs in nature. Proteins have different shapes to match the job they do. Proteins are long chains of 20 different **amino acids**. Different combinations of amino acids give different proteins.

The order of the bases on DNA forms a code. This code carries the instructions for the order of amino acids in new proteins.

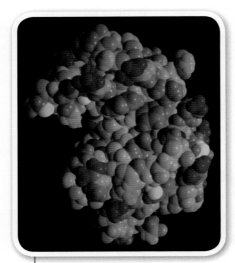

Enzymes work because of the shape of their active sites.

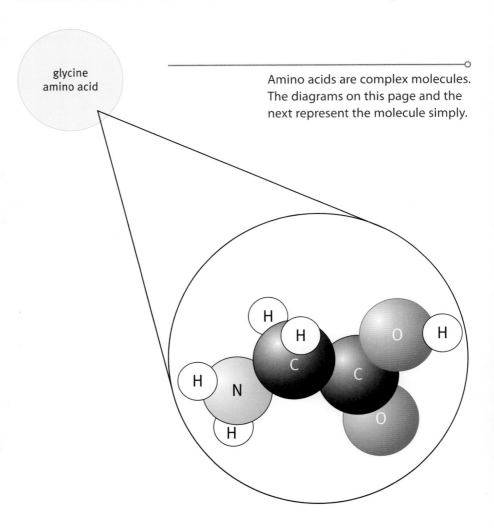

glycine amino acid

Amino acids are complex molecules. The diagrams on this page and the next represent the molecule simply.

Proteins are made in the cytoplasm

DNA contains the genetic code for making proteins. DNA is found in the nucleus.

Proteins are made in the cytoplasm. A copy of the gene is made, which carries the genetic code from the nucleus to the cytoplasm. It goes through pores in the nuclear membrane.

Summary box

- The large, spiral DNA molecule holds genetic information in our cells.
- DNA has four bases: A pairs with T, C pairs with G.
- The bases are like rungs on a ladder.
- DNA can reproduce itself exactly by base pairing.

B5: GROWTH AND DEVELOPMENT

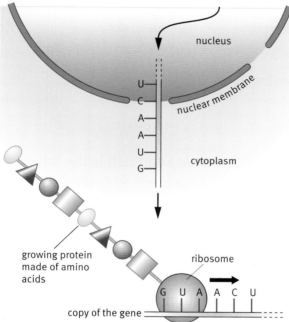

1. The gene unzips and a copy of the gene is made.

2. The copy of the gene moves out of the nucleus to the cytoplasm.

3. The copy of the gene spells out which protein to make.

The order of the bases on DNA is the genetic code for the production of a protein.

Questions

1. Why are instructions for making proteins copied onto a messenger molecule?
2. How does the order of bases on a DNA strand affect the proteins made in a cell?
3. Make bullet-point notes to explain how a protein is made. Your first bullet point should be:
   - The gene unzips.

   Your last bullet point should be:
   - The protein is released into the cytoplasm.

Summary box

- Proteins are made in the cytoplasm.
- The order of bases on the DNA spells out which protein to make.
- A copy of the gene carries the instructions in the DNA out of the nucleus.

H: MAKING PROTEINS

I Specialised cells – special proteins

Find out about

- some different proteins in your body
- why cells become specialised

An oak tree has about 30 different types of cell. Your body has more than 300 types of cell. Each cell type has its own set of proteins.

Different proteins for different jobs

Some proteins make up the framework of cells and tissues. These are **structural proteins**. If we take away all the water in an animal cell, 90% of the rest is proteins.

Protein	Found in …	Property
keratin	hair, nails, skin	strong and waterproof
elastin	skin	springy
collagen	skin, bone, tendons, ligaments	tough and not very stretchy

Different structural proteins have different properties.

The flesh of meat and fish is the animals' muscles. Muscles are mainly protein and provide the protein in many people's diets. Plant seeds such as soya and other beans are rich in proteins and are the basis of many vegetarian meals.

Other proteins are essential for the chemical reactions that keep our bodies working. For example, **enzymes** speed up the chemical reactions in a cell. **Antibodies** are the proteins that help to defend us against disease.

An oak tree has about 30 cell types.

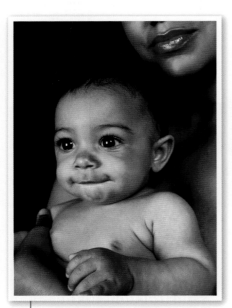

Humans have over 300 different types of cell.

Questions

1. Name three types of protein in your body.
2. Name one structural protein and say how it is suited to do its job.

Different cells have different genes switched on

The genes in a cell are made up of DNA. Each gene is the instruction for a cell to make a different protein. By controlling what proteins a cell makes, genes control how a cell develops.

Each of your cells has a copy of all your genes. Something must make some cells turn into nerve cells and others into heart cells and all the other cell types. There must be **genetic switches**, but how they work is still being researched by scientists.

> **Summary box**
> - There are lots of different types of proteins including structural proteins, antibodies, and enzymes.
> - Different cells have different genes switched on.
> - Our bodies have over 300 different cell types.

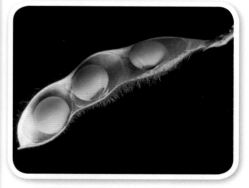

Rhino horn, tortoise shell, soya beans, steak ... lots of different proteins.

Questions

3 DNA contains a cell's instructions. What does DNA do in a cell?

4 How do genes control how a cell develops?

Genes can be switched on and off

Each gene codes for one type of protein. So, in any living thing, there are as many genes as there are different types of protein. In humans there are 20 000 to 25 000 genes. Not all these genes are active in every cell. As cells grow and specialise, some genes switch off.

In a hair cell, the genes for the enzymes that make the hair protein keratin will be switched on:

hair cell genes switched on → enzymes for making keratin → hair grows

In a salivary gland cell:

salivary gland cell genes switched on → amylase secreted → starch digested

Different genes switch off as an embryo grows

An early embryo is made entirely of embryonic stem cells. These cells are unspecialised up to the eight-cell stage. These stem cells can develop into any kind of cell. All the genes in these cells are switched on. As the embryo develops, cells specialise. Different genes switch off in different cells.

This is a chromosome. The green areas are the active genes where DNA unravels whilst the protein is being made.

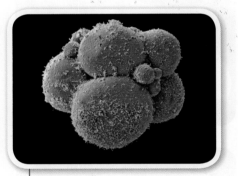

Up to the eight-cell stage, each cell of a human embryo can develop into any kind of cell – or even into a whole organism.

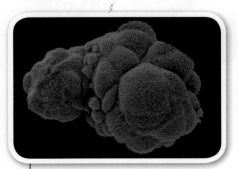

At six days and about 50 cells, the cells in this human embryo are already specialised and will form different types of tissue. The different cells can no longer develop into a whole organism.

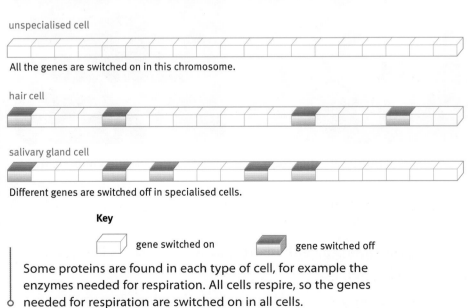

Some proteins are found in each type of cell, for example the enzymes needed for respiration. All cells respire, so the genes needed for respiration are switched on in all cells.

In adults, there are stem cells too. They are found in parts of the body where of worn-out cells need replacing. These stem cells can only develop into specialised replacement cells. So some of their genes must be switched off.

The right genes are switched on in the right part of the body

Compare the fingers on your right hand with the same fingers on your left hand. They are probably almost mirror images of each other. As we grow, the position and type of cells must be controlled, so each tissue and organ develops in the right place.

Cells near the end of a limb will make fingers. Cells nearer the body will make the arm. This happens because of the difference in the concentrations of chemical signals in each region of the embryo.

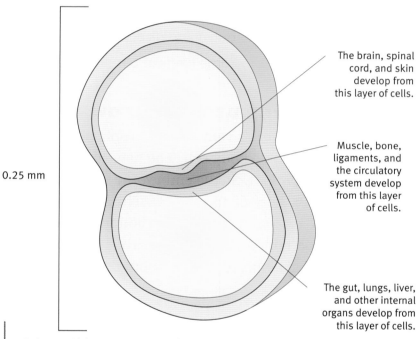

0.25 mm

The brain, spinal cord, and skin develop from this layer of cells.

Muscle, bone, ligaments, and the circulatory system develop from this layer of cells.

The gut, lungs, liver, and other internal organs develop from this layer of cells.

It is possible to map specialised parts of the body onto a diagram of a 14-day-old human embryo. Here you can see which groups of cells in the embryo will develop into future tissues and organs.

Questions

5 Suggest a function, other than respiration, that all cells carry out.

6 At the eight-cell stage of any embryo, how many genes are switched on?

7 What is the evidence that some genes are switched off at the 50–100-cell stage?

Summary box
- The right genes need to be switched on in the right cell.
- Up to the eight-cell stage the early embryo:
  - is made of embryonic stem cells
  - has every gene switched on
  - can develop into any kind of cell.
- Adult stem cells only make specialised replacement cells.

Stem cells

Find out about

- scientific research to use stem cells for treating some diseases

A lot of research is going on into stem cells. This is because scientists think stem cells could be used to:
- treat some diseases
- replace damaged tissue.

Imagine if scientists could produce …	They might use them to treat …
nerve cells	Parkinson's disease and spinal cord injuries
heart muscle cells	damage caused by a heart attack
insulin-secreting cells	diabetes
skin cells	burns and ulcers
retina cells	some kinds of blindness

Therapeutic cloning using your own cells

The nucleus from an egg cell is removed and replaced with the nucleus from a patient's own body cell. The new embryo has the same genes as the patient. So the embryonic stem cells produced would match those of the patient. The cells could be transplanted into the patient to make them better. There would be no problem of rejection. This is **therapeutic cloning**. Some people do not agree that cloned embryos should be produced.

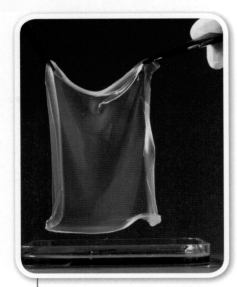

Scientists grew this skin from skin stem cells in sterile conditions. Doctors use it for skin grafts.

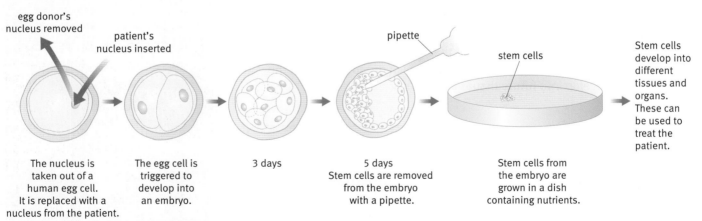

The nucleus is taken out of a human egg cell. It is replaced with a nucleus from the patient.

The egg cell is triggered to develop into an embryo.

3 days

5 days
Stem cells are removed from the embryo with a pipette.

Stem cells from the embryo are grown in a dish containing nutrients.

Stem cells develop into different tissues and organs. These can be used to treat the patient.

Therapeutic cloning is controversial because it uses stem cells from an embryo.

Using adult stem cells

Doctors can avoid using embryos by using the 'adult' stem cells already in your body. Adult bone marrow stem cells are already used to treat leukaemia (cancer of the blood). The donated stem cells restore healthy blood cell production in the patient.

More useful 'adult' stem cells can be collected from the placenta once a child is born. Some companies will store your loose milk teeth in a special freezer so that the stem cells they contain could be used to treat you later. These stem cells can grow into a range of specialised cells.

Some companies will store your baby teeth. The stem cells that they contain might help to treat you years later.

This patient received a new windpipe. The cartilage had been made using stem cells collected from her own bone marrow. There was no risk of rejection.

It might be possible to use a patient's own stem cells to replace any of their damaged cells. There wouldn't be a risk of rejection and the cells might restore organs that could not otherwise heal themselves.

Some successes using stem cell treatments

Government regulatory bodies have not approved any treatments yet for widespread use, but trials and experiments are underway. Doctors need to be sure that the cells will behave in the right way when they are implanted into the body.

Questions

1. Why is therapeutic cloning using embryonic stem cells controversial?
2. Describe how stem cells from a donor could treat leukaemia (blood cancer).
3. Name two potential sources of adult stem cells.
4. What are the advantages of using stem cells from your own body to treat an illness?

Summary box
- Stem cells can grow into replacement cells and tissues.
- Stem cells may treat illnesses or replace damaged tissue without rejection.
- Therapeutic cloning is controversial because it uses stem cells from an embryo.
- 'Adult' stem cells for treatment can be collected from bone marrow, placenta, and milk teeth.

Science Explanations

Genetic technologies are at the cutting edge of modern science. The study of proteins made in cells, stem cell technology, and cellular control are key areas of research. These fields promise to provide huge benefits to present and future generations.

You should know:

- that in multicellular organisms, cells are often specialised to do a particular job
- that up to the eight-cell stage, all the cells in a human embryo are identical (embryonic stem cells) and can produce any type of human cell
- why some cells (adult stem cells) remain unspecialised, becoming specialised at a later stage
- that in plants, only cells within special regions called meristems undergo cell division – these can develop into any kind of plant cell and are used to produce clones
- about the range of tissues that unspecialised plant cells can become
- that plant hormones cause cut stems from a plant to develop roots and grow into a complete plant, which is a clone of the parent
- that the environment affects growth and development of plants, for example, plants grow towards light (phototropism), and how this increases the plant's chance of survival
- how genetic information is stored within cells
- why cell division by mitosis produces two new cells identical to each other and to the parent cell
- what happens during the main processes of the cell cycle (cell growth and mitosis)
- why a special type of cell division called meiosis is needed to produce gametes
- that instructions for making proteins are coded for by genes in the nucleus, but proteins are produced in the cell cytoplasm
- that four different bases make up both strands of the DNA molecule and that these always pair up in the same way, A with T and C with G
- that the order of bases in a gene is the genetic code for the production of a protein
- why body cells in an organism contain the same genes, but they only produce the specific proteins they need because the genes that are not needed are switched off
- how any gene can be switched on during development in embryonic stem cells to produce any type of specialised cell
- the applications for which adult stem cells and embryonic stem cells can potentially be used.

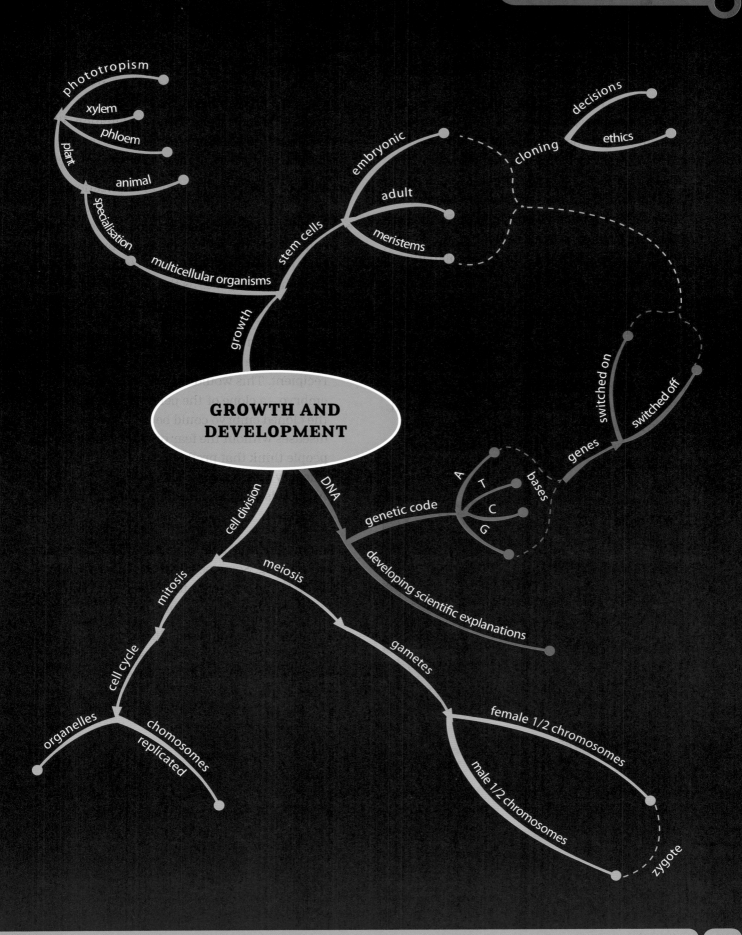

Ideas about Science

In this module you learn more about growth and development. You should become more aware that scientists need to work out explanations of their observations and measurements.

- Watson and Crick discovered the structure of DNA in 1953. They looked at all the evidence gathered in the previous 50 years and produced a model that matched the evidence. It suggested that the DNA structure is a double spiral (a double helix). Developing this explanation required creative thought.

You need to distinguish between data and the explanations that explain the data. You also need to suggest whether the data and the explanation agree or not. You need to recognise when creative thought is involved in producing an explanation.

- Watson and Crick's DNA structure accounted for what was known about DNA so far. It also let them predict how DNA might copy itself exactly.

Scientific explanations allow predictions to be made. If the prediction proves to be correct, this increases the confidence in the explanation. When a prediction proves to be wrong, either the prediction or the explanation may be incorrect.

The use of stem cells has ethical implications.

- The potential of stem cells to treat diseases and mend damaged tissue is enormous. Some stem cells come from human embryos, but there is a high risk of rejection if these stem cells do not have the same genes as the person getting the transplant. There are also ethical issues about using embryos in this way.
- One solution is to replace the nucleus of a zygote with a nucleus from a body cell of the recipient. This would then develop into an embryo – a clone of the person. Stem cells from the cloned embryo could be used to treat diseases without the fear of rejection. Some people think that producing clones of people in this way is wrong.
- All of this work is subject to government regulation.

It is sometimes difficult to decide what is right and what is wrong. Some people think that the right decision is one that leads to the best outcome for the greatest number of people involved.

Other people think that certain actions are either right or wrong whatever the consequences. You need to identify arguments that are based on these two different ideas.

Review Questions

B5: GROWTH AND DEVELOPMENT

1 **a** Copy the table below and put a tick in the box next to the correct answer. What job do genes do in our cells?

store glucose	
describe how to make proteins	
release energy by respiration	
transport materials around a cell	

b Here are some statements about our genes.

Copy the table below and put a tick in the box in each row to show if the statement is true or false.

	true	false
Genes are found on the chromosomes.		
Chromosomes are found in the cell cytoplasm.		
There is one gene on each chromosome.		
Genes are made from proteins.		

2 DNA is a vital molecule.
Copy the table and put a ring around the correct answer in each row. What are the features of DNA?

DNA feature				
number of strands	1	2	3	4
number of different bases	2	3	4	5
arrangement of bases between strands	single	pairs	triplets	fours
shape of molecule	circular	cubic	spiral (helix)	zigzag

3 Meiosis is a special type of cell division.
Meiosis makes egg and sperm cells (gametes).

When an egg is fertilised by a sperm cell a zygote is formed. Four people are talking about chromosomes, meiosis, and fertilisation.

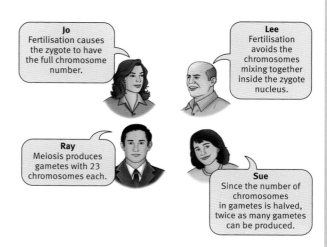

Jo Fertilisation causes the zygote to have the full chromosome number.

Lee Fertilisation avoids the chromosomes mixing together inside the zygote nucleus.

Ray Meiosis produces gametes with 23 chromosomes each.

Sue Since the number of chromosomes in gametes is halved, twice as many gametes can be produced.

Two people are correct. Two people's ideas are wrong. Which two people are right?

4 Sperm and egg cells meet at fertilisation. At first when the cells divide they are all the same – they are unspecialised. However, soon the cells of the embryo become specialised. Think about these facts about the growing embryo.

Copy the table and put a tick in the box in each row to show if these statements are true or false.

	true	false
The embryo cells no longer contain the same genes.		
Some of the embryo genes are no longer active.		
Each embryo cell produces only the specific proteins it needs.		
There are laws that control experiments on embryos.		

C5 Chemicals of the natural environment

Why study chemicals of the natural environment?

Conditions on Earth are special. The temperature is just right for most water to be liquid. The atmosphere has enough oxygen for living things to breathe, but not so much that everything catches fire. Rocks are the source of many of the chemicals that meet our daily needs.

What you already know

- Bonds between atoms in molecules are strong, while forces between molecules are very weak.
- An atom consists of a central nucleus surrounded by electrons.
- Ionic compounds conduct electricity when molten or in solution, because the ions are free to move.
- During a chemical reaction, atoms are conserved.
- A chemical change caused by the flow of an electric current is called electrolysis.
- Making, using, and disposing of products can affect the environment.

Find out about

- the chemicals in spheres of the Earth
- theories of structure and bonding
- tests for analysing water quality
- methods of extracting metals from ores.

The Science

Theories of structure and bonding explain how atoms are arranged and held together in all the chemicals that make up our Earth. There are three types of strong bonding: metallic, ionic, and covalent. There are weaker forces of attraction between molecules.

Ideas about Science

The properties of metals make them very useful. But mining, mineral processing, and metal extraction can all have a serious impact on the environment. Benefits must be weighed against costs.

A Chemicals in spheres

Find out about

- **naturally occurring elements and compounds**
- **cycling of elements and compounds between the spheres**

People who study the Earth often think of it as being made up of spheres (see the diagram below). Starting from the middle, first comes the core, then the mantle, then the crust.

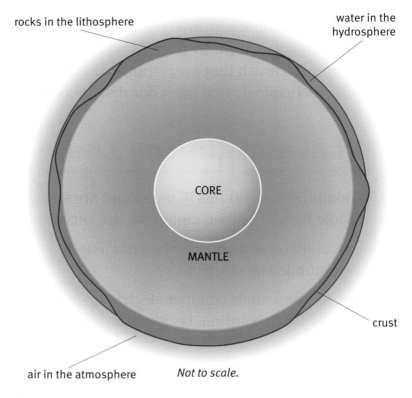

The Earth's spheres. The outer spheres are much thinner than this relative to the core and mantle.

This module looks at the spheres that make up the Earth's surface and provide us with natural resources: the lithosphere, hydrosphere, and atmosphere. We obtain all we need from the Earth.

- The **lithosphere** is the **crust** and upper **mantle**. The lithosphere is about 100 km thick.
- The oceans and rivers make up the **hydrosphere**.
- The **atmosphere** is a layer of air around the Earth.

There are living things in parts of all three outer spheres.

Question

1 'Living things inhabit the atmosphere, hydrosphere, and lithosphere.' Give two examples of things that live in each sphere.

Elements in the spheres

The spheres vary greatly in their chemical composition.
- The atmosphere consists mainly of two uncombined elements: nitrogen and oxygen.
- The hydrosphere is almost entirely the compound water.

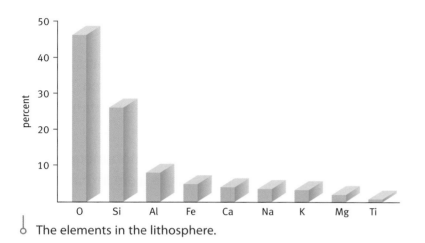

The elements in the lithosphere.

- The rocks of the lithosphere are mainly silicates. These are compounds of silicon combined with oxygen and small quantities of other elements.

Elements such as carbon and oxygen exist as different chemical species in each of the three spheres.

Flowing between the spheres

Chemicals do not always stay in one sphere. Some are constantly on the move between the spheres. Think of a carbon atom: it may start in the atmosphere; be taken into a plant; be washed into water in the hydrosphere; then buried in sediment of the lithosphere.

Water flows freely between the spheres. The obvious place for water is the hydrosphere. But think of clouds in the atmosphere, then rain sinking into the lithosphere, reappearing as a spring, from where it flows into a river and then into the sea.

Summary box
- The Earth is made up of the core, mantle, and crust.
- We obtain all we need from the Earth.
- The lithosphere, hydrosphere, and atmosphere provide us with natural resources.
- Chemicals move between the Earth's spheres.

Questions

2 Look at the bar chart on this page. What are the three most abundant elements in the lithosphere?

3 Make a list of four different things you use in a typical day that came from the lithosphere.

B | Chemicals of the atmosphere

Find out about

- gases in the air
- weak attractions between molecules
- strong covalent bonding

nitrogen	N N
oxygen	O O
argon	Ar
carbon dioxide	O C O

The main atmospheric gases.

Questions

1 Copy and complete:

Gas in air	Formula
nitrogen	
oxygen	
argon	
	CO_2

2 From the table of atmospheric gases, name:
   a an element made of single atoms
   b two elements made of molecules
   c a compound made of molecules.

The Earth is just the right size for its gravity to hold onto its atmosphere. It is also just the right distance from the Sun to have the right temperature for water to exist as a liquid. This water, together with the carbon dioxide and oxygen in the atmosphere, allows the Earth to support a great variety of plant and animal life.

The main gases in the atmosphere are in these proportions: 78% nitrogen (N_2), 21% oxygen (O_2), 1% argon (Ar), and 0.04% carbon dioxide (CO_2). The atmosphere also contains water vapour and small amounts of other gases.

All the chemicals in the atmosphere are either non-metallic elements (eg, O_2, N_2, Ar) or compounds made from atoms of non-metallic elements.

Atoms and molecules in the air

Most of the chemicals in the atmosphere are made of **small molecules**. A few are single atoms. These are noble gases from Group 0 of the periodic table.

All molecules have a slight tendency to stick together. For example, there is an attraction between one O_2 and another O_2 molecule. But these **attractive forces** are very weak; this is why the chemicals that make up the atmosphere are gases with low melting and boiling points.

One way to picture this is that the molecules are moving so quickly that, when two O_2 molecules come close to each other, the attractive force between them is not strong enough to hold them together.

Strong bonds in molecules

The bonds inside molecules that hold the atoms together are called **covalent** bonds. These bonds are very strong, many times stronger than the weak attractions between molecules. Small molecules such as O_2 or H_2 do not split up into atoms except at very, very high temperatures.

Chemists often use a single line to represent a single bond between atoms in a molecule. For example, the simple molecules in hydrogen, oxygen, and carbon dioxide can be represented by the molecular formulae H_2, O_2, and CO_2. But if you want to show their bonds, they can be represented by:

$$H-H \qquad O=O \qquad O=C=O$$

Some **molecular models** use the same idea. A coloured ball represents each atom, and a stick or a spring is used for each bond joining them together. Different elements form different numbers of covalent bonds. A hydrogen atom will form just one bond, while an oxygen atom will form two.

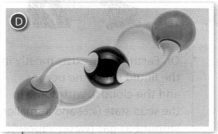

Ball-and-stick models of molecules.
White = hydrogen, H.
Black = carbon, C.
Red = oxygen, O.

Questions

3 Which of the images (A–D) on this page represents:
   a CO_2?   b O_2?   c H_2O?   d H_2?

4 The molecular formula for oxygen is O_2. Write the molecular formula for the following.
   a N≡N
   b Cl—Cl
   c
   $$\begin{array}{c} H \\ | \\ H-C-H \\ | \\ H \end{array}$$

5 Nitrogen makes three covalent bonds. Fluorine, chlorine, and hydrogen each make one convalent bond. Draw diagrams to show the covalent bonding in these molecules:
   a fluorine, F_2
   b hydrogen chloride, HCl
   c ammonia, NH_3.

6 Explain why the gases in the atmosphere have low melting and boiling points. Use ideas about attractive forces in your answer.

7 Estimate the volume of argon in the room in which you are sitting.

C Chemicals of the hydrosphere

Find out about

- unusual properties of water
- ions in solution

Properties of water

We see water so often that we take its special chemical and physical properties for granted.

One of these special properties is that it is a liquid at room temperature even though it is made up of small molecules. O_2, N_2, and CO_2 are all gases at room temperature. So you might expect H_2O to be a gas at room temperature too. But H_2O melts at 0 °C and boils at 100 °C. This is because the attractive forces between water molecules are slightly stronger than the very weak attractive forces between molecules of O_2, N_2 and CO_2.

Another special property of water is that it is a good solvent for **salts**. Most common solvents do not **dissolve** ionic compounds, but water does.

Pure water does not conduct electricity – in this way it is like other liquids made up of small molecules. This shows that it does not contain charged particles that are free to move.

Even though pure water does not conduct, you should never touch electrical devices with wet hands. This is because the water on your hands is not pure, so it will conduct electricity and increases the risk of you receiving an electric shock.

On Earth, water exists mostly in the liquid state (the oceans and the clouds), with some in the solid state (ice) and a smaller amount as gas (water vapour).

Molecule	Boiling point (°C)
H_2O	100
N_2	−196
O_2	−183
CO_2	−78

The boiling point of some of the main atmospheric gases. Water has an unusually high boiling point for a molecule of its size.

Questions

1 Write down the sentence from this page that explains the unusual boiling point of water.

2 How is water different to most common solvents?

3 Carbon monoxide is a small molecule similar to N_2, O_2, and CO_2. Which of the following temperatures is most likely to be its boiling point?
  a 236 °C
  b 94 °C
  c −191 °C.

Why is sea water salty?

The diagram below shows how soluble chemicals get carried from rocks to the sea during part of the water cycle.

The main soluble chemical carried into the sea is sodium chloride (common salt). River water does not taste salty because the concentration is so low, but the concentration of salt in the sea has built up over millions of years, and so it does taste salty.

> **Summary box**
> - Water is unusual for a small molecule because it is a liquid and dissolves salts.
> - When salts dissolve, the ions separate and can move freely.

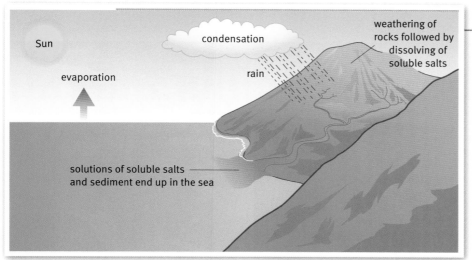

Weathering slowly breaks down rocks. Rainwater falls on the broken rocks. Soluble chemicals in the rocks dissolve in the water and get washed away.

Other chemicals that dissolve include potassium chloride, magnesium chloride, and sodium sulfate. One litre of typical seawater contains about 40 g of dissolved chemicals from rocks.

Most of the compounds that are dissolved in seawater are made up of positively charged metal ions and negatively charged non-metal ions. They are salts.

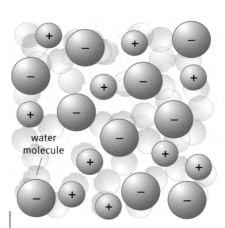

Ions separate and can move about freely when dissolved in water. In seawater there is a mixture of positive ions and negative ions.

Questions

4 Why does the sea taste salty while river water does not?

5 When a salt dissolves in water, what happens to the ions?

D | Detecting ions in salts

Find out about

- **properties of ions**
- **precipitation reactions**
- **ionic equations**
- **tests for ions**

Why test water?

- Water coming into at homes has to meet government standards.
- There are rules about which chemicals industrial companies are allowed to release into rivers.
- Beaches are tested for the quality of the seawater.

There are good reasons for all of us to be interested in the make up of Earth's hydrosphere.

Water at home.

A beach in Devon.

Industrial discharge to a river.

Ions

Simple tests can help us find out about the ions in a sample of water. These tests rely on each type of ion from the dissolved salts having special, unique properties.

Sodium chloride has a unique set of properties, such as melting point and solubility. But some of its properties are shared by all salts containing sodium ions. Some are shared by all salts containing chloride ions.

Question

1 Which is more soluble, sodium chloride or calcium carbonate?

Solubility

Solubility in water is an important property of ionic compounds. Sodium chloride is soluble in water over a wide range of temperatures. So a solution may contain quite high concentrations of Na^+ and Cl^- ions together.

Calcium carbonate, however, has only a low solubility, so no solution may contain high concentrations of calcium ions (Ca^{2+}) and carbonate ions (CO_3^{2-}) ions together. You can see this if you mix together two solutions: one with a high concentration of Ca^{2+} ions, the other with a high concentration of CO_3^{2-} ions. Calcium nitrate and sodium carbonate would be good examples to use.

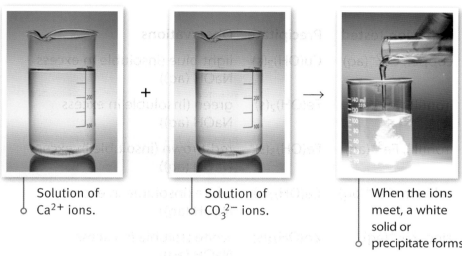

| Solution of Ca^{2+} ions. | Solution of CO_3^{2-} ions. | When the ions meet, a white solid or precipitate forms. | Ions of an insoluble salt cluster together and make a solid precipitate. |

When the Ca^{2+} ions meet the CO_3^{2-} ions, they cluster together to make solid crystals, which fall to the bottom of the solution. This is called a **precipitate**. It is exactly the reverse of dissolving.

Using precipitation reactions

Water companies use precipitation reactions to check the concentration of ions in drinking water. They make sure the water is safe to drink and will not cause too much limescale.

Some metal ions can be toxic if the concentration is too high, while high levels of sulfate ions can have a laxative effect. High levels of magnesium ions and calcium ions can cause too much limescale.

Questions

2 What is the name for a solid formed when two solutions mix?

3 What would happen if you mixed solutions of potassium carbonate and calcium nitrate?

The hydroxide precipitates $Fe(OH)_2$, $Fe(OH)_3$, $Cu(OH)_2$.

Looking for metal ions

Several metal ion solutions form a solid precipitate when a solution of hydroxide ions is added. This is because the metal hydroxide compounds are insoluble.

Many of these hydroxide compounds are coloured. This means that the colours can be used to identify the metal ion in a solution.

Metal ion tested	Precipitate	Observations
copper, $Cu^{2+}(aq)$	$Cu(OH)_2(s)$	light blue (insoluble in excess NaOH (aq))
iron(II), $Fe^{2+}(aq)$	$Fe(OH)_2(s)$	green (insoluble in excess NaOH (aq))
iron(III), $Fe^{3+}(aq)$	$Fe(OH)_3(s)$	red-brown (insoluble in excess NaOH (aq))
calcium, $Ca^{2+}(aq)$	$Ca(OH)_2(s)$	white (insoluble in excess NaOH (aq))
zinc, $Zn^{2+}(aq)$	$Zn(OH)_2(s)$	white (soluble in excess NaOH (aq))

The colours of hydroxide precipitates.

To test for a metal ion, add a few drops of a dilute solution of sodium hydroxide to the solution you want to test.

If a solid precipitate forms, the table above can be used to find out which metal ions are in the solution.

In the test for zinc ions, if you add excess sodium hydroxide solution the white precipitate dissolves, leaving a colourless solution. The other hydroxide precipitates in the table do not dissolve when you add excess sodium hydroxide.

Questions

4 A few drops of dilute sodium hydroxide are added to a colourless solution. A red–brown precipitate forms. Which metal ion is present?

5 A solution gives a green precipitate when dilute sodium hydroxide is added. Which metal ion is present?

Looking for non-metal ions

Some tests for non-metal ions also use precipitation reactions.

Chloride, bromide, iodide

Silver nitrate solution is used to test for chloride (Cl⁻), bromide (Br⁻), and iodide (I⁻) ions. They form precipitates with distinct colours: chloride ions form a white precipitate, bromide ions form a cream precipitate, and iodide ions form a yellow precipitate.

Sulfate

Barium nitrate or barium chloride solutions are used to test for sulfate ions (SO_4^{2-}). A white precipitate forms.

The precipitates AgCl(s), AgBr(s), and AgI(s).

Ion tested	Test	Precipitate	Colour of precipitate
Cl⁻(aq)	acidify with dilute nitric acid, then add silver nitrate solution	AgCl(s)	white
Br⁻(aq)	acidify with dilute nitric acid, then add silver nitrate solution	AgBr(s)	cream
I⁻(aq)	acidify with dilute nitric acid, then add silver nitrate solution	AgI(s)	yellow
SO_4^{2-} (aq)	acidify, then add barium chloride solution or barium nitrate solution	BaSO₄(s)	white

The colours of precipitates used for testing non-metal ions.

Carbonate

Carbonate ions (CO_3^{2-}) make carbon dioxide gas when you add a dilute acid. Bubbles form, creating effervescence, and the gas can be tested to see if it makes limewater go cloudy.

The test for carbonate ions. Bubbles of carbon dioxide form with a dilute acid.

Questions

6 What would you expect to see if you added a solution of barium nitrate to an acidified solution of sodium sulfate?

7 A colourless solution gives a yellow precipitate when silver nitrate is added to the acidified solution. Which negative ion does it contain?

Summary box

- Some ionic compounds are soluble, others are insoluble.
- A precipitate is a solid formed when two solutions mix.
- Different colours of precipitates can be used to identify ions.

E — Chemicals of the lithosphere

Find out about

- chemicals from the Earth's crust
- ionic bonding
- giant ionic structures

Rocks and minerals

The lithosphere is made of **rocks**. Rocks can be big (mountains), medium-sized (boulders), or small (stones and pebbles).

Rocks may contain one or more **minerals**. Minerals are naturally occurring chemicals; they are often compounds but may be elements.

Limestone rock is made mainly from the mineral calcite, a compound with the chemical formula $CaCO_3$.

Gold is found naturally on its own in the lithosphere. Gold is very unreactive so it does not usually combine with other elements to form compounds.

The non-metals oxygen and silicon are the two most **abundant**, or common, elements in the lithosphere. They form silica and silicate minerals, which make up 95% of the Earth's continental crust. Quartz is a form of silica, with silicon and oxygen atoms joined by covalent bonds.

Aluminium is the third most abundant element in the lithosphere. It is found in a mineral called feldspar, which also contains silicon and oxygen.

Minerals from seawater

Seawater contains dissolved chemicals. When the water evaporates, **ionic compounds** crystallise. Sodium chloride, NaCl, or rock salt, is a common example.

Calcite, $CaCO_3$.

Granite is a mixture of minerals.

In every crystal of sodium chloride, the ions are arranged in the same regular pattern. All crystals of the compound have the same cubic shape.

Questions

1. Write a sentence that makes clear the differences between the words *rock*, *mineral*, and *lithosphere*.
2. What are the three most common elements in the lithosphere?

The structure and properties of salts

Crystals of sodium chloride are cubes. They are made up of sodium and chloride ions.

When a sodium chloride crystal forms, millions of Na^+ ions and millions of Cl^- ions pack closely together. The ions are held together very strongly by the attraction between their opposite charges. This is called **ionic bonding**, and the structure is called a **giant ionic lattice**.

Unlike compounds such as water, which are made up of individual molecules of H_2O, there is *not* an individual NaCl molecule.

Because of the very strong attractive forces, it takes a lot of energy to break down the regular arrangement of ions. So NaCl has to be heated to 801°C before it melts, and to 1413 °C before it boils.

Compare this with the melting point of ice (0°C) and the boiling point of water (100 °C).

To melt ice and boil water you only need to break the weak forces of attraction *between* the water molecules.

To conduct electricity, a chemical must have charged particles that can move. Ionic solids have charged particles, but they cannot move, so they do not conduct. A melted or dissolved ionic compound *will* conduct electricity because the ions are free to move about.

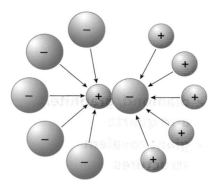

The Na^+ ions attract Cl^- ions that are close to it. The Cl^- ions attract Na^+ ions that are close to it.

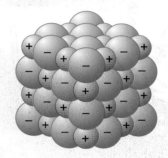

Each of these ions then attracts other ions of the opposite charge, and the process continues until millions of oppositely charged ions are all packed closely together.

How a sodium chloride crystal forms.

Questions

3 Decide whether each of the following statements about ionic compounds such as sodium chloride are true or false. If the statement is false, rewrite it so that it is true.
   a Ionic compounds have low melting points and boiling points.
   b Ionic solids conduct electricity.
   c Ionic compounds conduct eletricity when melted or dissolved in water.

4 Write a sentence to explain each of the properties in question 3. For example, ionic compounds have (high/low) melting and boiling points because …

Summary box
- The lithosphere is made of rocks. Rocks may contain minerals, which are naturally occuring chemicals.
- Ions that have opposite charges are attracted to each other. This is called ionic bonding.
- A giant ionic lattice is formed when millions of ions join together to make a solid.

F Carbon minerals – hard and soft

Find out about

- diamond, graphite, and quartz
- giant covalent structures

Diamond and graphite – non-identical twins

If you heat a **diamond** strongly, it will eventually burn and make carbon dioxide. **Graphite** will do the same. This is because both these minerals are made from carbon and nothing else.

The carbon atoms in diamond and graphite are arranged and stuck to each other in different ways. Diamond and graphite are good examples of how the behaviour of a solid depends on its structure and bonding.

Diamond and graphite head-to-head

Property	Diamond	Graphite
hardness (1 = softest, 10 = hardest)	10	1–2
melting point (°C)	3560	3650
boiling point (°C)	4830	4830
solubility in water	insoluble	insoluble
electrical conductivity	low	high

Some properties of diamond and graphite.

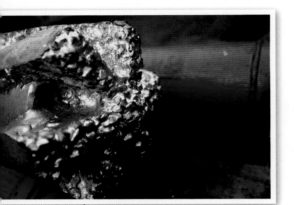

A diamond-tipped drill. The extreme hardness of diamond allows it to cut through anything.

Carbon atoms in diamond

In diamond, covalent bonds join each carbon atom to its four nearest neighbours. The four bonds are evenly spaced in *three dimensions* around each carbon atom, making pyramid shapes.

The arrangement is called a **giant covalent structure**. It repeats over and over again until you have billions and billions of atoms covalently bonded together. The covalent bonds are all very strong.

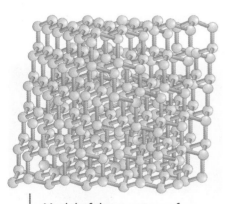

Model of the structure of diamond. Each carbon atom inside the structure is covalently bonded to four others in a huge 3D arrangement.

Carbon atoms in graphite

Graphite has a giant covalent structure, too. Covalent bonds join each carbon atom to its three nearest neighbours. The three bonds are evenly spaced in *two dimensions*, making flat sheets or layers of hexagons.

These layers include billions of atoms. Each atom has one electron that has not been used to make a covalent bond.

These electrons drift around in the gaps between the layers of atoms. They help stick the layers together, but only weakly.

Different structures – different properties – different uses

Diamond and graphite both have high melting points and boiling points. This is because they have lots of strong covalent bonds. These need to break before the solid can melt. This takes lots of energy so the melting point is high. Diamond and graphite are also both insoluble in water for similar reasons.

Strong covalent bonds in all directions make diamond the hardest known material. Drills used in mining have bits with diamond tips.

In graphite, weak forces between the layers make it easy for them to slide over one another. This slipperiness makes graphite useful as a lubricant.

The electrons drifting between the layers in graphite are free to move. This allows graphite to conduct electricity. In contrast, diamond has no charged particles that are free to move, and it does not conduct.

Silicon dioxide

Silicon dioxide has a giant covalent structure similar to diamond. The strong bonds between the silicon and oxygen atoms make it strong and rigid. It is hard, so it is suitable for use in sandpaper as an abrasive.

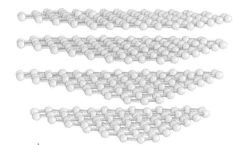

Model of the structure of graphite. There are flat sheets of hexagons with weak forces between them.

Pure silicon dioxide crystals are transparent and very hard.

Summary box
- Diamond and graphite are both minerals made of carbon atoms.
- They have giant covalent structures.
- They have different properties because the atoms are held together in different ways.
- Silicon dioxide is another giant covalent compound.

Questions

1. Explain why diamond and graphite have high melting points.
2. Why does graphite conduct electricity but diamond does not?
3. Predict whether silicon dioxide will:
   a. have a high or low melting point
   b. conduct electricity
   c. dissolve in water.

G | Metals from the lithosphere

Find out about

- metal ores
- extracting metals
- electrolysis

Metal ores

The wealth of societies has often depended on their ability to extract and use metals. Mining and quarrying for metal ores takes place on a large scale and can make large profits. Many people are employed by the mining industry but its processes can have a major impact on the environment.

All metals come from the lithosphere, but most metals are too reactive to exist on their own in the ground. Instead, they exist combined with other elements as compounds. Like other compounds found in the lithosphere, they are called minerals.

Rocks that contain useful minerals from which metals can be extracted are called **ores**. The valuable minerals are often the **oxides** or sulfides of metals.

Gold is unusual because it occurs on its own. It has been used by humans for more than 5000 years. More **reactive metals** like iron were not used by humans until methods for **extracting** them from their ores had been developed.

Extraction methods include **reduction** reactions. These separate the metal from the other elements it is combined with.

Mineral processing

Over hundreds of millions of years, rich deposits of ores have built up in certain parts of the Earth's crust. But even the richest deposits do not contain pure mineral. The valuable mineral is mixed with lots of useless dirt and rock, which have to be separated off as much as possible, creating lots of waste. This is called concentrating the ore.

Some ores are fairly concentrated when they are dug up. Other ores are much less concentrated – copper ore usually contains less than 1% of the pure copper mineral.

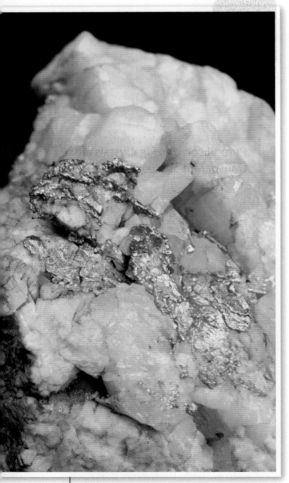

Gold is so unreactive that it occurs uncombined in the lithosphere. Most metals occur as compounds.

Question

1. A new mine is planned near a small town with high unemployment. Give one benefit and one cost for the people living in the town.

Extracting metals: some of the issues

There is a range of factors to weigh up when thinking about the method for extracting a metal.

- How can the ore be reduced?

The more reactive the metal, the harder it is to reduce its ore. The table on the right compares the methods used to reduce different ores.

- Is there a good supply of ore?

Metals ores are mined in different parts of the world. If ore is not very pure, it may not be worth using. The more valuable the metal, the lower the quality of ore that can be used.

- What are the energy costs?

It takes energy to extract metals. This is especially true if the metal is extracted by electrolysis. For example, a quarter of the cost of making aluminium is the cost of electricity.

- What is the impact on the environment?

Metals like iron and aluminium are produced on a huge scale. Millions of tonnes of ore are needed. Mining this ore can have a big environmental impact. This is why it is important to recycle metals. It takes about 250 kg of copper ore to make 1 kg of copper. So recycling 1 kg of copper means that 250 kg of ore does not need to be dug up.

Metal	Method
potassium, sodium, calcium, magnesium, aluminium (MORE REACTIVE)	electrolysis of molten ores
zinc, iron, tin, lead, copper	reduction of ores using carbon
silver, gold (LESS REACTIVE)	metals occur uncombined

Question

2 Suggest explanations for these facts:
  a The Romans used copper, iron, and gold, but not aluminium.
  b Iron is cheap compared with many other metals.
  c Gold is expensive, even though it is found uncombined in nature.
  d About half the iron we use is recycled, but nearly all the gold is recycled.
  e The tin mines in Cornwall have closed, even though there is still some tin ore left in the ground.

An open-pit copper mine in Utah, USA. Mining on this scale has a big impact on the environment.

Extracting metals from ores

Zinc is a metal that can be extracted from its oxide. The task is to remove the oxygen from the zinc, to convert zinc oxide (ZnO) to zinc (Zn). Because oxygen is removed this is called reduction.

Carbon is used to take the oxygen away. The carbon is **oxidised**.

zinc loses oxygen to carbon and is reduced…

zinc oxide + carbon ⟶ zinc + carbon monoxide

…carbon takes oxygen from zinc and is oxidised

Reducing zinc oxide to zinc using carbon.

This reaction can also be written as a symbol equation. Atoms are not created or destroyed during reactions. They are simply rearranged.

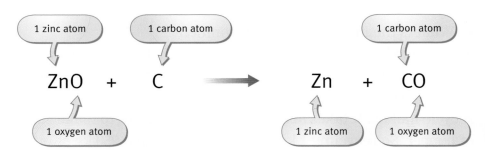

Further oxidation of the carbon forms carbon dioxide.

Carbon is often used to extract metals. Carbon, in the form of coke, can be made cheaply from coal. At high temperatures, carbon has a strong tendency to react with oxygen, so it is good at removing oxygen from metals.

Iron and copper can also be extracted with carbon. These reactions can be summarised as:

iron oxide + carbon ⟶ iron + carbon dioxide

copper oxide + carbon ⟶ copper + carbon dioxide

Hot metal can be poured.

Questions

3 Copy and complete. Words to use: gains, oxidised, oxygen, loses. When a metal oxide _____ oxygen it is reduced. The carbon _____ this _____ and is _____.

4 The formula iron oxide is Fe_2O_3.
  a How many iron atoms are there in the iron oxide formula?
  b How many oxygen atoms are there in the iron oxide formula?

Relative atomic masses

Chemists need to know the relative masses of atoms to answer questions such as 'how much iron is there in iron oxide?'

Atoms are far too small to weigh directly. Instead of working in grams, chemists find the mass of atoms relative to one another. This is called the **relative atomic mass**.

Values for the relative atomic masses of elements are shown in the periodic table in C4. The relative mass of the lightest atom, hydrogen, is 1.

One Mg atom weighs twice as much as one C atom

Formula masses

If you know the formula of a compound, you can work out its **relative formula mass** by adding up the relative atomic masses of all the atoms in the formula:

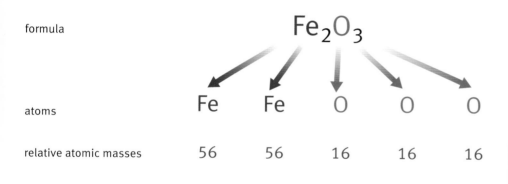

Finding the formula mass of Fe_2O_3.

The gram formula mass of a compound is an amount equal to the relative formula mass in grams. So the gram formula mass of Fe_2O_3 is 160 g.

> **Worked example**
>
> What is the mass of iron in the gram formula mass of Fe_2O_3?
>
> Relative formula mass of Fe_2O_3 = 160
>
> Fe_2O_3 has two Fe atoms
>
> The mass of Fe in 160 g of Fe_2O_3 = 56 × 2 = 112 g

Questions

Look up relative atomic masses in the periodic table in C4.

5 What are the relative atomic masses of:
   a C?
   b O?
   c Al?
   d Cu?
   e Ca?

6 What is the relative formula mass of carbon dioxide?

7 What is the relative formula mass of:
   a Al_2O_3?
   b CuO?
   c $CaCO_3$?

8 What is the mass of aluminium in the gram formula mass of Al_2O_3?

Extracting aluminium

Some reactive metals, such as aluminium, hold on to oxygen so strongly that they cannot be extracted using carbon. To extract these metals, **electrolysis** has to be used.

Aluminium is the most abundant metal in the lithosphere, but it is very hard to separate the aluminium from its ore. Aluminium ore consists mainly of aluminium oxide, Al_2O_3.

The diagram below shows the equipment used to extract aluminium by electrolysis. The process takes place in steel tanks lined with carbon. The carbon lining is the negative **electrode** and conducts electricity.

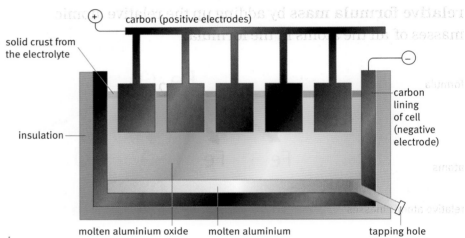

Equipment for extracting aluminium from its oxide by electrolysis.

The **electrolyte** is hot, molten aluminium oxide. It contains Al^{3+} and O^{2-} ions. The aluminium ions are positively charged and are attracted to the negative electrode lining the tank. Aluminium forms at the negative electrode. Because it is very hot, the aluminium is a liquid and forms a pool of molten metal at the bottom of the tank.

The positive electrodes are blocks of carbon dipping into the molten oxide. The oxide ions are negatively charged and are attracted to the positive electrodes. Oxygen forms at the positive electrodes. Some of this combines with the carbon to make carbon dioxide.

Worker in an aluminium processing plant removing bars of aluminium from their moulds. The aluminium leaving the electrolysis tanks is molten (melted). It is poured into moulds and cools to form solid metal bars.

Questions

9 Why can aluminium not be extracted from its ore by heating with carbon?

10 An aluminium atom has 13 electrons. Draw a diagram to show the electron arrangements in:
   a an aluminium atom
   b an aluminium ion.

Electrolysis

When an ionic crystal melts, the ions are no longer held tightly together. They are free to move. Because it contains charged particles (ions) that can move about, the molten ionic compound will conduct electricity.

Molten ionic compounds are electrolytes. They can be decomposed or turned back into their elements by passing electricity through them.

During electrolysis:
- metals form at the negative electrode
- non-metals form at the positive electrode.

Aluminium is extracted from aluminium oxide by electrolysis. This takes place in huge tanks.

Questions

11 In the extraction of aluminium by electrolysis, molten aluminium oxide is used as the electrolyte. Why is it important that the electrolyte is molten?

12 Sodium chloride can be melted and electrolysed. What will form at:
   a the positive electrode?
   b the negative electrode?

Summary box
- Metals are extracted from rocks called ores.
- Metals that are not very reactive can be extracted from their oxides by heating with carbon. The carbon is oxidised and the metal oxide is reduced.
- More reactive metals such as aluminium are extracted by electrolysis.
- During electrolysis metals form at the negative electrode, and non-metals form at the positive electrode.

H | Structure and bonding in metals

Find out about

- the properties of metals
- the structure of metals

Metal properties

Metals have been part of human history for thousands of years. Our lives still depend on metals, even though there are lots of other new materials to choose from, including all the different plastics. The properties of metals are shown by the ways in which they are used.

Most metals have **high melting points**.

Many metals are **strong**. The titanium hull of this research submarine needs to be strong. Titanium is also used to make hip joints and racing cars.

Metals can be bent or pressed into shape. They bend without breaking. They are **malleable**. Aluminium sheet can be moulded under pressure to make cans.

Metals **conduct electricity**. Copper and aluminium are commonly used as conductors.

Metallic structures

Scientists can collect data about the properties of metals. For example, they can measure the temperature at which a metal melts. Metals have high melting points but to explain *why* metals have high melting points scientists need to know something about their structure.

Questions

1. List four properties of metals. Use the pictures above to help you.
2. Give three examples of metals used for their strength. For each metal, give an example of a use that depends on the strength of the metal.

Scientists use models to describe what they have discovered about the structures of metals. The models show that metals are made up of atoms that are:
- tiny spheres
- arranged in a regular pattern
- packed close together as a giant structure.

This model was thought up creatively to explain the properties of metals. A model is a good one if it can explain all the data.

The diagram on the right shows the arrangement of atoms in copper, a typical metal. You can see how closely together the atoms of copper are packed – as close together as it is possible to be.

The arrangement of atoms in copper.

Metallic bonding

Metals have a special kind of bonding – not ionic, nor covalent, but metallic. The atoms are held to each other by strong **metallic bonding**. Because the bonding is strong, copper is strong and difficult to melt. It takes lots of energy to break the bonding, separate the atoms, and turn the solid metal into a liquid, so the melting point is high.

Questions

3 Someone looking at a model showing the arrangement of atoms in a copper crystal might think that the following statements are true. Which of these ideas are true and which are false? Re-write the false statements to make them true.
  a Copper is dense because the atoms are closely packed.
  b Copper has a high melting point because the atoms are strongly bonded in a giant structure.
  c Copper melts when strongly heated because the atoms melt.

4 Give two examples of products that are made of metals because they have high melting points.

Summary box
- Metals are strong, malleable, conduct electricity, and have high melting points.
- Metals atoms are held close together by strong metallic bonding.

I The life cycle of metals

Find out about
- impacts of extracting, using, and disposing of metals

Mining, mineral processing, and metal extraction produce many valuable metal products. These activities can also have a serious impact on the environment. There can be a conflict between those who want to build up profitable industries and those whose aim is to protect the natural world.

Mining

Mines can be on the surface (open-cast) or underground. Both types of mine produce large amounts of waste rock and can leave very large holes in the ground.

Miners working in underground mines are at risk from dust, heat, and the possibility of the mine caving in. Modern mines have ventilation shafts and fans to provide fresh air into the mine and sensors to monitor the air quality. Walls and ceilings are braced with strong boards to prevent cave-ins.

Mining is much safer now than it has been in the past but accidents still happen. It is impossible to make mining completely safe but most countries now have regulations to ensure that mining companies reduce risks.

Processing ores

Many metal ores do not contain much of the metal. The ore in an open-pit copper mine may contain as little as 0.4% of the metal. This means that nearly all of the rock dug from the ground becomes waste.

Near any mine there are waste tips. These can be a hazard if they contain small amounts of toxic metals such as lead or mercury.

Metal extraction

All the stages of metal extraction need energy, use large volumes of water, and give off air pollutants. Industries now have to do more to prevent harmful chemicals escaping into the environment. Companies try to develop equipment and procedures that minimise the use of energy, water, and other resources.

A large pond in Jamaica used to contain the waste from an aluminium mine. The main impurity is iron oxide, which ends up in the rusty-looking waste.

Summary box
- Extracting, using, and disposing of metals all impact on the environment.
- Recycling metals helps to reduce the amounts of raw material, energy, and water used.

Metals in use

Careful choice of metals can reduce the environmental impact of our lifestyle. In transport, for example, lighter cars, trucks, and trains mean less fuel is used and emissions are reduced. Vehicles can be designed to be lighter by replacing steel with lighter metals such as aluminium, or with plastics.

Recycling

Recycling is well established in the metal industries. Scrap metal from all stages of production is routinely recycled. Much metal is also recycled at the end of the useful life of metal products.

Recycled steel is as good as new after reprocessing. The scrap is fed to a furnace and melted with fresh metal to make new steel. For every tonne of steel recycled, there is a saving of 1.5 tonnes of iron ore and half a tonne of coal.

Recycling aluminium is particularly cost-effective because so much energy is needed to extract the metal by electrolysis.

Recycling reduces the impact on the environment by cutting the use of raw materials, and the associated mining and processing.

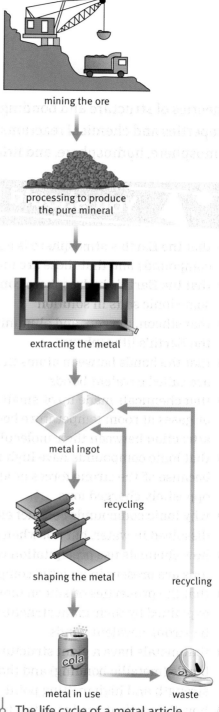

The life cycle of a metal article.

Questions

1 How can mining and processing ores affect the environment?
2 Give one risk arising from mining metal ores, and one way in which this risk is reduced.
3 Why is using lighter metals, like aluminium, in cars better for the environment?
4 Why is recycling aluminium particularly cost effective?
5 Known reserves of aluminium ore (bauxite) will last for hundreds of years. The aluminium industry claims this makes aluminium a sustainable material. Do you agree?

Science Explanations

Theories of structure and bonding can help to explain the physical properties and chemical reactions of the chemicals we find in the atmosphere, hydrosphere, and lithosphere.

You should know:

- that the Earth's atmosphere is a mixture of elements and compounds and that these are made up of small molecules
- that the Earth's hydrosphere consists mainly of water with some ionic salts in solution
- that silicon, oxygen, and aluminium are very abundant in the Earth's lithosphere
- that the bonds between atoms in a molecule are strong and are called covalent bonds
- that chemicals made up of small molecules are often liquids or gases at room temperature because of the weak forces of attraction between their molecules
- that ionic compounds have high melting and boiling points because of the strong forces of attraction between oppositely charged ions
- why ionic compounds conduct electricity when molten or dissolved in water, but not when solid
- how chemists use precipitation reactions to detect which ions are present in an ionic compound
- that the properties of silicon dioxide and diamond can be explained by their giant structures of atoms held together by strong covalent bonds
- that metals have a giant structure held together by strong bonds (metallic bonding) and that this explains their strength and high melting point
- how the method used to extract a metal from its ore is related to the reactivity of the metal
- how to calculate the mass of the metal that can be extracted from a metal compound
- that when a metal loses oxygen it is reduced, while carbon gains oxygen and is oxidised
- that during electrolysis metals form at the negative electrode and non-metals form at the positive electrode.

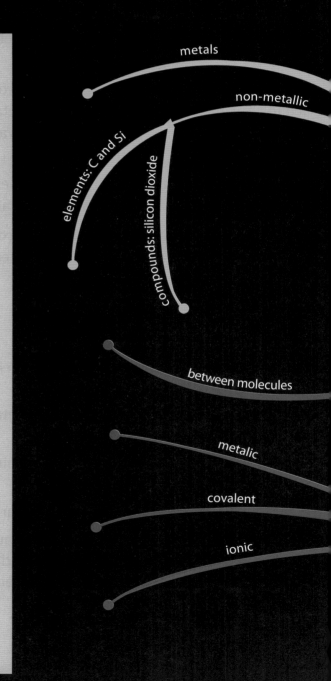

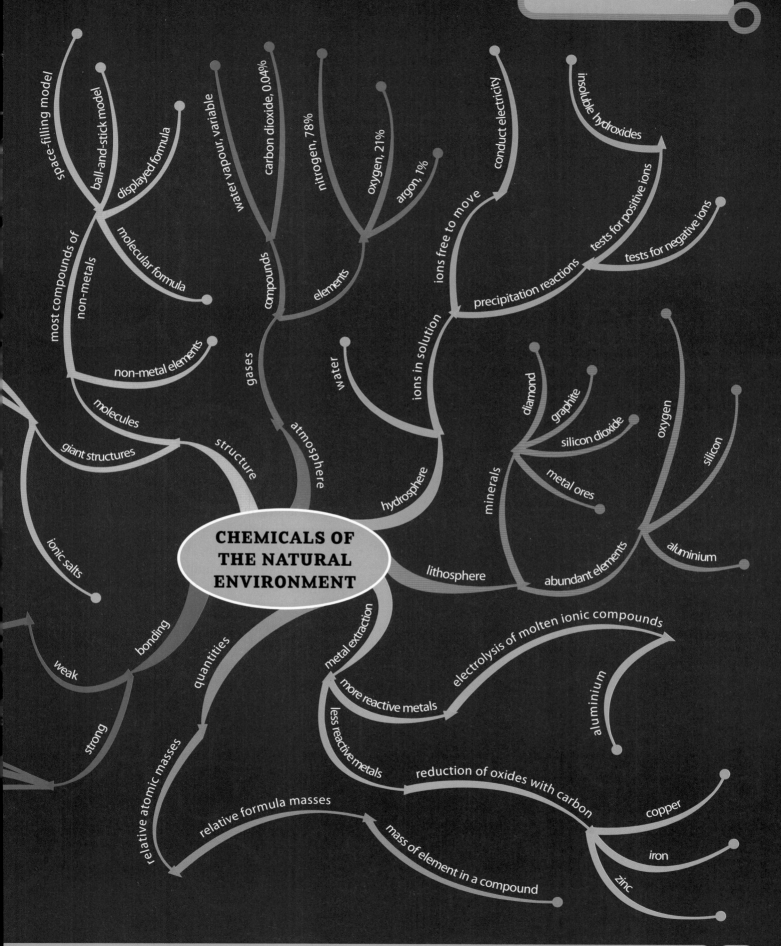

Ideas about Science

Scientific explanations are based on data but they go beyond the data. An explanation has to be thought up creatively to account for the data.

In the context of the theories of structure and bonding you should be able to:
- identify statements that report information collected (data) and statements of explanatory ideas (hypotheses, explanations, and theories) in an account of scientific work
- recognise data, such as measures of the properties of elements and compounds, that is explained by theories of structure and bonding.

New technologies and processes based on scientific advances sometimes introduce new risks. Some people are worried about risks associated with mining and metal extraction.

You should be able to:
- explain why nothing is completely safe
- identify examples of risks that arise from mining and metal extraction
- suggest ways of reducing a given risk.

Some applications of science, such as the extraction and use of metals, can have unintended and undesirable impacts on the quality of life or the environment. Benefits need to be weighed against costs. You should be able to:
- identify the groups of people affected, and the main benefits and costs of a course of action for each group
- identify and suggest examples of unintended impacts of human activity on the environment, such as the production of large volumes of waste by mining, mineral processing, and metal extraction based on low-grade ores
- explain the idea of sustainability, and apply it to the methods used to obtain, use, recycle, and dispose of metals.

Some forms of scientific work have ethical implications that some people will agree with and others will not. When an ethical issue is involved, you need to be able to:
- state clearly what the issue is
- summarise the different views that people might hold.

When discussing ethical issues, common arguments are that:
- the right decision is the one that leads to the best outcome for the majority of the people involved
- certain actions are right or wrong whatever the consequences.

Review Questions

C5: CHEMICALS IN NATURE

1 A chemist tests a solution of an impure solid. His results are in the table. Use pages 146 and 147 to help you answer the questions.

Test number	Test	Observation
1	acidify, then add silver nitrate solution	white precipitate
2	acidify, then add barium nitrate solution	white precipitate
3	add dilute sodium hydroxide solution	light-blue precipitate

a Which test shows that the solid contains copper ions?
b Which test shows that the solid contains chloride ions?

2 Zinc metal can be extracted from its oxide by heating with carbon. The equation for the reaction is:

$$ZnO + C \longrightarrow Zn + CO$$

a In the equation above:
  i name the element that is oxidised
  ii name the compound that is reduced.
b Calculate:
  i the formula mass of zinc oxide (relative atomic masses: Zn=65, O=16, C=12)
  ii the mass of zinc in 81 kg of zinc oxide
  iii the mass of zinc in 1 kg of zinc oxide.
c Explain why recycling helps to make the use of zinc metal more sustainable.

3 Aluminium metal is extracted from its oxide by electrolysis. The electrolyte is molten aluminium oxide, Al_2O_3.
a Explain why aluminium cannot be extracted from its oxide by heating with carbon.
b Explain why molten aluminium oxide conducts electricity while solid aluminium oxide does not.
c During the electrolysis of molten aluminium oxide what forms at the:
  i positive electrode?
  ii negative electrode?

4 The table gives the melting point and solubility of three different chemicals.

Chemical	Melting point (°C)	Solubility in water
nitrogen	−210	insoluble
potassium chloride	770	soluble
silicon dioxide (quartz)	1610	insoluble

a For each chemical in the table above state:
  i the type of bonding
  ii the type of structure.
b Use ideas about structure and bonding to explain the difference in melting point between nitrogen and potassium chloride.
c Why is silicon dioxide found in the lithosphere and not in the hydrosphere or the atmosphere?

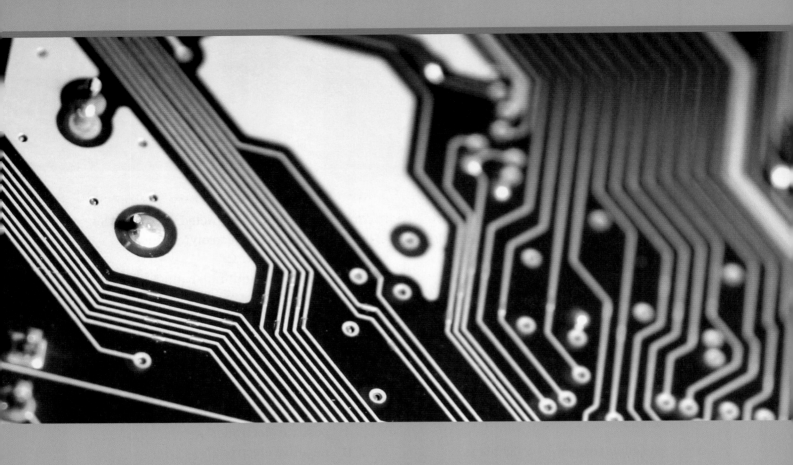

P5 Electric circuits

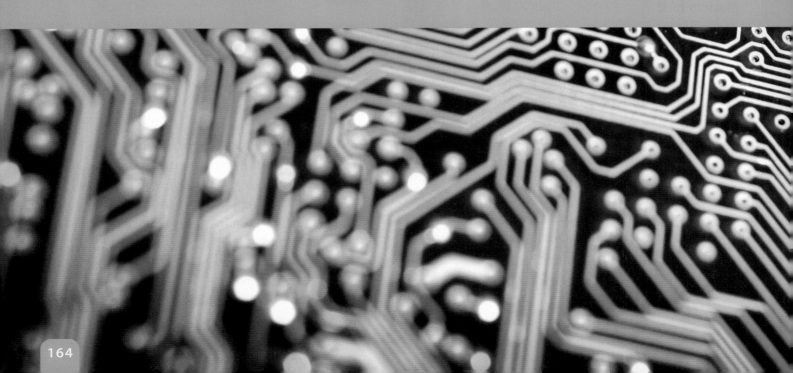

P5: ELECTRIC CIRCUITS

Why study electric circuits?

Imagine life without electricity – rooms lit by candles or oil lamps, no electric cookers or kettles, no television, computers, or mobile phones, no cars or aeroplanes. Electricity has transformed our lives, but you need to know enough to use it safely.

What you already know

- Electric current is not used up in a circuit.
- Electric current transfers energy from the battery to components in the circuit.
- The power of a device is measured in watts and power = voltage × current.
- There is a magnetic field near a wire carrying a current. This can be used to make an electromagnet.
- In a power station a turbine drives a generator to produce electricity.

Find out about

- the idea of electric charge, and how moving charges result in an electric current
- electric current, voltage, and resistance
- energy transfers in electric circuits, and how mains electricity is generated and distributed
- electric motors.

The Science

Particles in atoms carry an electric charge. An electric current is a flow of charges. The size of current depends on the voltage and the resistance of the circuit. A voltage can be produced by moving a magnet near a coil.

Physics in Action

Scientists use electricity, or instruments that depend on electricity, in almost every aspect of their work. One important focus of research in the 21st century is on the development of new ways of generating electricity. Another is to develop better electric motors for cars.

A Electric charge

Find out about

- electric charge
- the effects of like and unlike charges on each other
- the connection between charge and electric current

When you get out of a car you sometimes get a small electric shock when you touch the car door. A small spark jumps between your hand and the car door. When you take off a jumper you sometimes hear little crackles. In a dark room you may sometimes see the sparks.

These effects are caused by electric charges moving. Electric charge is a property of matter. The photograph shows that lightening is a very dramatic effect of electric charges moving.

Charging by rubbing

Electrical effects can be produced by rubbing two materials together. If you rub a balloon against your jumper, the balloon will stick to a wall. If you rub a plastic comb on your sleeve, the comb will pick up small pieces of tissue paper.

When you rub a piece of plastic, it is changed so that it affects objects nearby. Electric charge is being stored on the plastic. If a lot is stored, it may escape by jumping to a nearby object, in the form of a spark. We say that the plastic has been charged. Charging by rubbing also explains the effects described above.

A lightning strike is electric charges moving at high speed from a thundercloud to the ground or vice versa.

Two types of charge

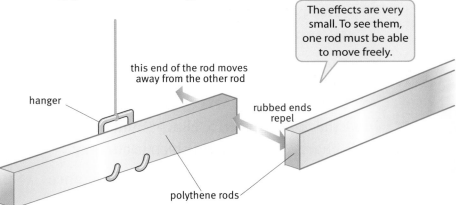

The effects are very small. To see them, one rod must be able to move freely.

this end of the rod moves away from the other rod

hanger

rubbed ends repel

polythene rods

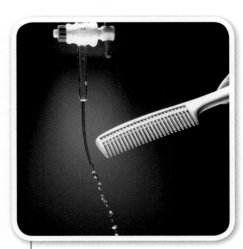

A charged comb attracts a stream of water.

Rub two rods of the same plastic and they repel. When you rub rods made of different plastics, you can find some pairs that **attract** each other. Scientists' explanation of this is that there are two types of **electric charge**, called **positive** and **negative**. These names are just labels. They could have called them red and blue, or A and B.

A science explanation: moving electrons

Scientists explain that this is because of the movement of small negatively charged particles called electrons.

This flow chart explains how an object is charged by rubbing.

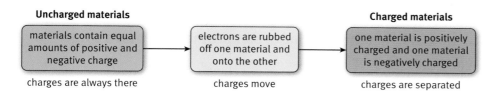

Some materials lose electrons when rubbed and the cloth gains electrons. Other materials gain electrons, so the cloth loses them.

All the materials charged by rubbing must be insulators; in conductors like metals electrons quickly move to even up the charges again.

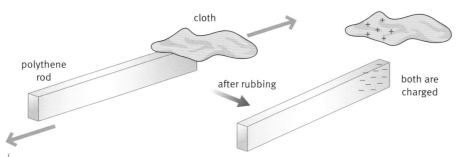

After it has been rubbed, the rod has a negative charge and the cloth has a positive charge. A possible explanation is that some electrons have been transferred from the cloth to the rod.

Static electricity

The effects discussed in this section are often called electrostatic effects, and are said to be due to **static electricity**. The word 'static' indicates that the charges do not move easily through the materials.

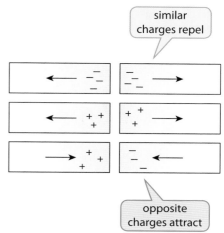

Why are light objects attracted to charged object?

A charged rod attract light unchanged objects, such as little pieces of paper. For example, when a negatively charged rod comes near the paper, the rod repels the negatively charged electrons in the paper to the end farthest away. This leaves the near end positively charged. So the little piece of paper is attracted to the rod.

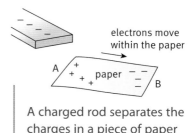

A charged rod separates the charges in a piece of paper and the paper is attracted to the rod.

Question

1 The diagram shows charged rods A, B, and C.

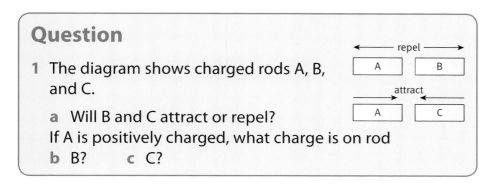

   a Will B and C attract or repel?

   If A is positively charged, what charge is on rod
   b B?   c C?

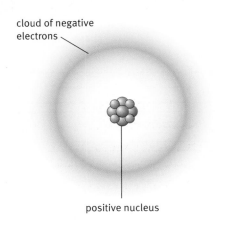

Atoms are made of	Electric charge
electrons	negative
protons	positive
neutrons	none, neutral

Summary box

- There are two types of electric charge: positive and negative.
- Like charges repel, opposite charges attract.
- Electrons are small, negatively charged particles.
- When materials gain electrons they become negatively charged.
- When they lose electrons they become positively charged.

What is electric charge?

Charge is a basic property of matter, which cannot be explained in terms of anything simpler. All matter is made of atoms, which are made out of protons, neutrons, and **electrons**. In most materials there are equal numbers of positive protons and negative electrons, so the whole thing is neutral. When you charge something, you move some electrons to it or from it.

The atom has a tiny positively charged nucleus. It is surrounded by a cloud of negatively charged electrons. As the electrons are on the outside of the atom they can be 'rubbed off', on to another object.

Moving charge = current

A van de Graaff generator is a machine for separating electric charge.

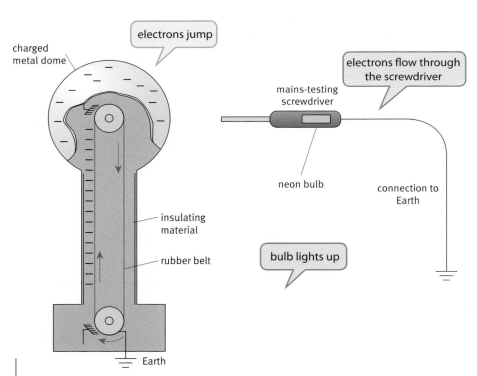

When charge flows from the dome, through the neon bulb, it makes it light up. This suggests that an electric current is a flow of charge.

Electric currents in circuits

A closed loop

The diagram below shows an **electric circuit**. You can use a circuit like this to show that:
- if you make a break *anywhere* in the circuit, *everything* stops
- both bulbs come on *immediately* when the circuit is completed
- they go off *immediately* if you make a break in the circuit.

One explanation for this is that there are tiny particles with electric charge in all the components of the circuit *all the time*. Closing the switch allows these charges to move through the wires, lamp filaments, and batteries. They all move together, so there is no delay, even if the charges move slowly.

A hamster and dried peas model

The diagram on the right shows a model that is useful for thinking about how a simple electric circuit works. The power source is the hamster turning the treadmill. As the treadmill turns, it pushes the peas along the pipe. The pipe is full of peas all the time, so they start moving everywhere around the circuit – immediately. The paddle wheel at the bottom turns as soon as the hamster runs. The hamster loses energy as it does work on the treadmill to make it turn and set the peas moving.

Find out about

- how simple electric circuits work
- models that help explain and predict the behaviour of electric circuits
- how to measure electric current

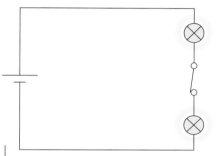

When you open the switch in this electric circuit, both bulbs go off.

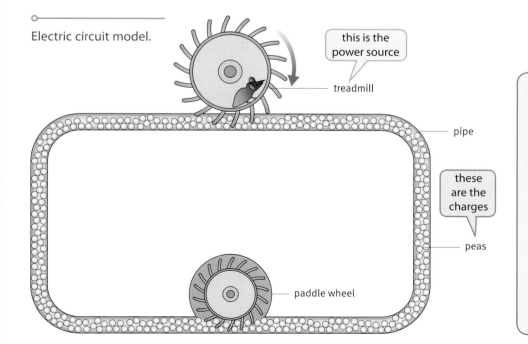

Electric circuit model.

Question

1. Look at the hamster electric circuit model. What corresponds to:
   a the battery?
   b the electric current?
   c Suggest one thing in a real electric circuit that might correspond to the paddle wheel.

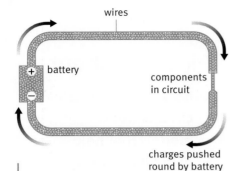

You can think of an electric current as a flow of charges.

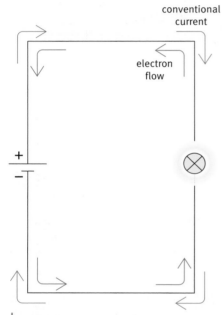

An electric current is a flow of electrons through the wires of the circuit. You can think of it equally well as a 'conventional current' of positive charges going the other way.

Conductors and insulators

Metals conduct electricity well because they have many electrons free to move. Insulators have very few electrons free to move.

An electric circuit model

A model is a way of thinking about how something works. The diagram on the left shows a useful scientific model of an electric circuit.

The key ideas are:
- charges are present throughout the circuit all the time
- when the circuit is a closed loop, the battery makes the charges move
- all of the charges move round together.

Chemical reactions inside the battery separate electric charges, so that positive charge collects on one terminal of the battery and negative charge on the other. When the battery is connected into a circuit, the charges on the battery terminals make free charges in the wire drift slowly along. All the charges begin to move at once, as soon as the battery is connected, so the effect of their motion is immediate. The flow of charge is continuous, all round the circuit. Charge also flows through the battery itself.

Conventional current, electron flow

In the model above, the charges move round the circuit away from the positive terminal of the battery, and back to the negative terminal of the battery. This assumes that the moving charges are positive. Long after this model was first proposed, scientists realised that metals are conductors because they contain lots of electrons that are free to move. Electrons have a negative charge.

So a current is a flow of negative electrons from the negative terminal round the circuit to the positive terminal.

In this course we use the model of conventional current going the other way to the flow of electrons.

Electric current

An electric current is a flow of charge. You cannot see a current, but you can observe its effects. The current through a torch bulb makes the fine wire of the filament heat up and glow.

To measure the size of an electric current we use an **ammeter**. The reading (in amperes, or amps (A) for short) indicates the amount of charge going through the ammeter every second.

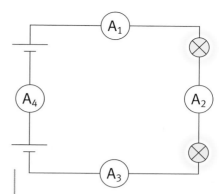

The current is the same size at all these points – even between the batteries. Current is not used up to make the bulbs light. This is a **series** circuit.

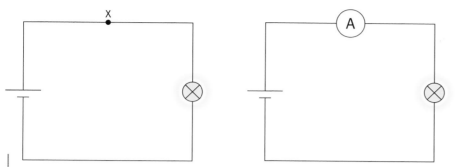

To measure the current at point X, you have to make a gap in the circuit at X and put the ammeter in the gap, so that the current flows through it.

Current around a series circuit

A series circuit has one single loop. When you use an ammeter to measure the electric current at different points around a series circuit, you find that:

- the current is the same everywhere in a single-loop electric circuit.

This may seem surprising. Current is not used up. Current is the movement of charges in the wire, all moving round together like dried peas in a tube. The current at every point round the circuit is the same.

Of course, *something* is being used up. It is the energy stored in the battery. This is getting less all the time. The battery is doing work to push the current through the filaments of the light bulbs, and this heats them up. The light then carries energy away from the glowing filament.

Summary box
- ✓ Electric current is a flow of charges.
- ✓ Electric current is measured in amps.
- ✓ The electric current through all the components in a series circuit is the same.
- ✓ In metals the charges that move are free electrons.

Question

2 How would you change the hamster electric circuit to model what happens in a series circuit with two identical bulbs? Use the model to explain:
   a why both bulbs go on and off together when the circuit is switched on and off
   b why both bulbs light immediately when the circuit is switched on
   c why both bulbs are equally bright.

C | Branching circuits

In this MP3 player, the battery runs the motor that turns the hard drive, the head that reads the disk, and the circuits that decode and amplify the signals.

Parallel circuits

In the circuit on the left there are two paths around the circuit. We say that the two bulbs are connected in **parallel**.

Advantages of parallel circuits:
- If one bulb burns out the other will stay lit.
- It is easy to spot a broken bulb and replace it.
- Components in parallel can be switched on and off independently.

Currents in the branches

Look at the circuit below, which has a motor and a buzzer connected in parallel. Nicola was asked to measure the current at points **a**, **b**, **c**, and **d**. Her results are shown in the table below on the right.

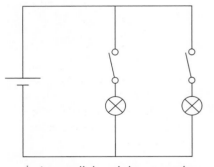

In parallel each lamp can be switched on and off independently.

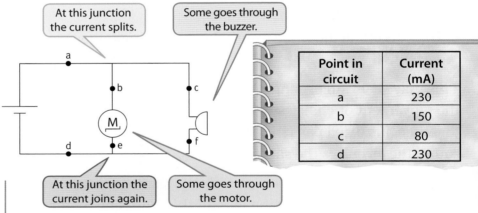

Point in circuit	Current (mA)
a	230
b	150
c	80
d	230

Measuring currents in a circuit with two parallel branches.

Nicola noticed that the current is the same size at points **a** and **d**: 230 mA. When she added the currents at **b** and **c**, the result was also 230 mA. The current in the two branches adds up to equal the total current in the single wire before the branching points, and after the branches meet again.

Summary box
- ✓ In a parallel circuit the total current from the battery is the sum of the currents through each of the components.
- ✓ In parallel, components can be switched on and off independently.

Question

1 Look at the circuit above with two lamps. Copy the diagram and add a switch that would turn both lamps on and off together.

Controlling the current

D

The size of the current depends on two factors:
- the voltage of the battery
- the resistance of the circuit components.

Battery voltage

Batteries come in different shapes and sizes. They usually have a **voltage**, measured in volts (V), marked on them, for example, 1.5 V, 4.5 V, 9 V.

> **Find out about**
> - the link between battery voltage, resistance, and current
> - what causes resistance

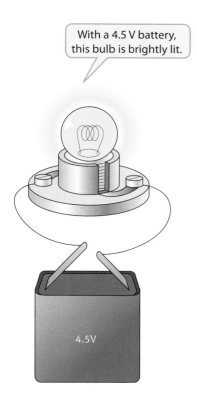

With a 4.5 V battery, this bulb is brightly lit.

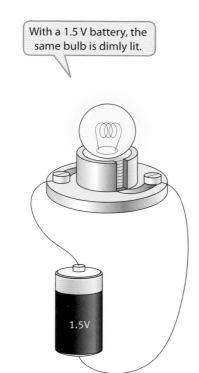

With a 1.5 V battery, the same bulb is dimly lit.

All the batteries on the front row are marked 1.5 V – but are very different sizes. The three at the back are marked 4.5 V, 6 V, and 9 V.

The bigger the current through a light bulb, the brighter it is. So the current through the bulb above is bigger with the 4.5 V battery. You can think of the voltage of a battery as a measure of its 'push' on the charges in the circuit. The bigger the voltage, the bigger the 'push' – and the bigger the current in the circuit.

The battery voltage depends on the choice of chemicals inside it. You can make a simple battery with two pieces of different metal and a beaker of salt solution or acid.

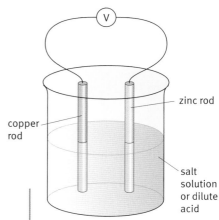

A simple battery. The voltage depends on the metals and the solution you choose.

D: CONTROLLING THE CURRENT 173

Resistance

The size of the current in a circuit depends on the resistance that the components in the circuit have to the flow of charge. The battery 'pushes' against this resistance. The diagram shows the effect of changing the resistance of a circuit. Resistors are components designed to control the flow of charge.

Changing resistance changes the size of the current. The bigger the resistance, the smaller the current.

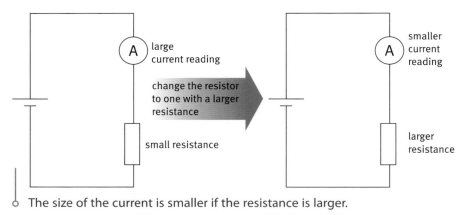

The size of the current is smaller if the resistance is larger.

A heating effect

Everything has resistance, not just special components called resistors. The resistance of connecting wires is very small, and can be ignored in calculations. Other kinds of metal wire have larger resistance. The filament of a light bulb has a lot of resistance. This is why it gets so hot and glows when there is a current through it. A heating element, like that in an electric kettle, is a resistor.

All metals get hot when charge flows through them.

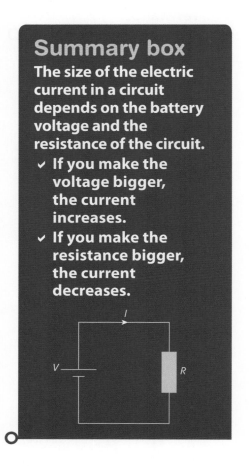

Summary box

The size of the electric current in a circuit depends on the battery voltage and the resistance of the circuit.

- ✓ If you make the voltage bigger, the current increases.
- ✓ If you make the resistance bigger, the current decreases.

Questions

1 Suggest two different ways in which you could change an electric circuit to make the electric current bigger.

2 Which has greater resistance, a connecting wire or the filament of a light bulb?

Measuring resistance

A student explores the relationship between current and voltage in a circuit.

The graph is a straight line because the resistance of the coil stays the same. So when you double the voltage you double the current. The number you get if you divide voltage by current is the same every time.

I connected a coil of resistance wire to a 1.5 V battery and an ammeter. She noted the current.

I then added a second battery in series and noted the current again.

I repeated this with 3, 4, 5, and 6 batteries, to get a set of results:

Number of 1.5 V batteries	Battery voltage (V)	Current (mA)
1	1.5	75
2	3.0	150
3	4.5	225
4	6.0	300
5	7.5	375
6	9.0	450

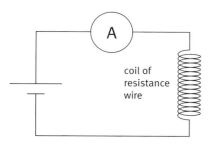
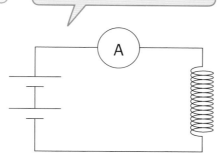

Finally, I drew a graph of current against battery voltage.

This is how to calculate resistance:

$$\text{resistance of a conductor} = \frac{\text{voltage across the conductor}}{\text{current through the conductor}}$$

The units of resistance are called ohms (Ω).

Worked example

A resistor is connected to a 9 V battery. The current through it is 30 mA. What is its resistance?

$$R = \frac{9V}{30mA} = \frac{9V}{0.03A} = 300\ \Omega$$

Questions

3 The voltage across a resistor is 12V. The current through it is 0.5 A. What is its resistance?

4 Use a pair of values from Keiko's results to work out the resistance of her coil in ohms.

5 In the circuit below, a 9 V battery is connected to a resistor. The reading on the ammeter is 0.2 A. What is the electrical resistance of the circuit?

Summary box

- Resistance = $\frac{\text{voltage}}{\text{current}}$
- Resistance is measured in ohms (Ω).
- A graph of current against voltage is a straight line when the resistance stays the same.

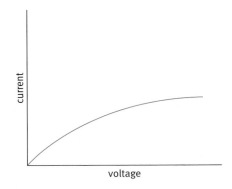

Changing resistance

Sometimes the resistance changes and then the current–voltage graph is not a straight line. This happens when a wire or a resistor gets hot and its resistance increases. A filament lamp is a good example. Changing the resistance of an electrical circuit is a useful way to control the current.

Resistance of a filament lamp

When current flows though a filament lamp it heats up and glows. Its resistance increases. It does not obey Ohm's law. The graph on the left, of current against voltage for a filament lamp, is curved.

Variable resistors

In an electric circuit sometimes we want to vary the current easily, for example, to change the volume on a radio or CD player. A variable resistor lets you do this. Its resistance can be steadily changed by turning a dial or moving a slider.

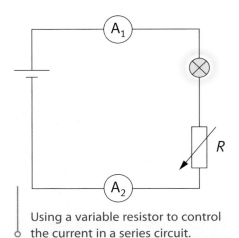

Using a variable resistor to control the current in a series circuit.

The circuit on the left shows a variable resistor connected to a lamp and a battery. As you increase the resistance of the variable resistor, the brightness of the light bulb decreases, and the readings on *both* ammeters decrease together. Increasing the resistance of the variable resistor decreases the size of the current everywhere round the circuit loop.

Each of these sliders adjusts the value of a variable resistor.

The light-dependent resistor (LDR)

A light-dependent resistor (LDR) is a component with a resistance that is large in the dark but gets smaller as the light falling on it gets brighter. An LDR is a sensing device that can be used to measure the brightness of light or to switch something on and off when the brightness of the light changes. For example, it could be used to switch an outdoor light on in the evening and off again in the morning.

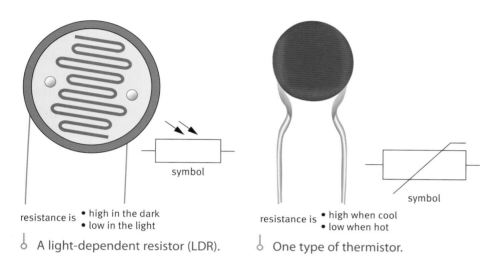

resistance is
- high in the dark
- low in the light

A light-dependent resistor (LDR).

resistance is
- high when cool
- low when hot

One type of thermistor.

The thermistor

A thermistor is a component with a resistance that changes with temperature. Its resistance decreases as its temperature increases. Thermistors can be used to make thermometers to measure temperature or to switch something on or off as temperature changes. For example, a thermistor could be used to switch a heater on when the temperature drops below a certain value and switch it off again when the temperature is back to the set value.

Questions

6 At night it gets dark and cold. What happens to the resistance of:
   a an LDR?
   b a thermistor?

7 A circuit contains an LDR. What happens to the current through the LDR as it gets darker?

Summary box

✓ Conductors and components that get hot, for example, filament lamps, have a constant resistance.
✓ Some useful components are:
  - a variable resistor – the resistance can be changed by moving a slider
  - a light-dependent resistor (LDR) – the resistance decreases as the light increases
  - a thermistor with a resistance that decreases as the temperature increases.

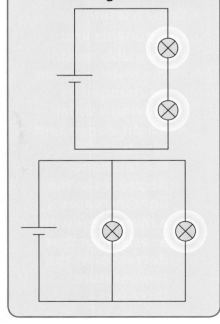

This circuit board from a computer contains many components.

Combinations of resistors

There are just two ways of connecting circuit components: in series or in parallel.

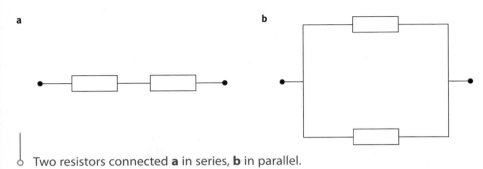

Two resistors connected **a** in series, **b** in parallel.

Two resistors in series have a larger resistance than one on its own. The battery has to push the charges through both of them.

Question

8 The light bulbs and batteries in these circuits are identical.
  a Which circuit has the largest total resistance?
  b Which circuit has the brightest bulbs?

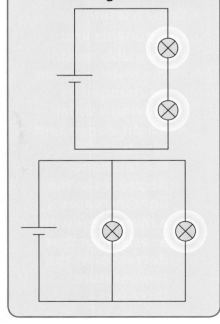

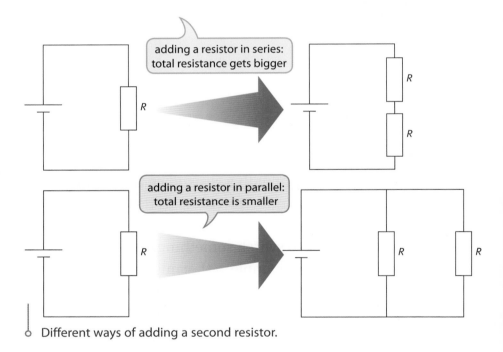

Different ways of adding a second resistor.

But connecting two resistors in parallel makes a smaller total resistance. There are now two paths that the moving charges can follow. Adding a second resistor in parallel does not affect the original path but adds a second one. It is now easier for the battery to push charges round, so the resistance is less.

Potential difference

E

A voltmeter shows a reading if you connect it across a resistor or bulb in a circuit like the one on the right. Resistors and bulbs do not 'push'. So the **voltmeter** must be showing something else.

The diagram below compares a water circuit to an electric circuit. The battery is like the pump lifting water to a higher level.

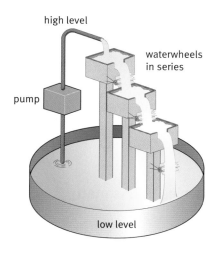

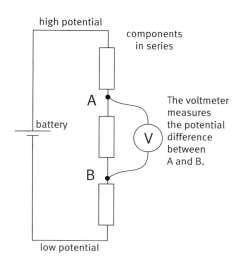

The voltmeter measures the potential difference between A and B.

Find out about

- ✓ **how voltmeters measure the potential difference between two points in a circuit**
- ✓ **how potential dividers are used**
- ✓ **how current splits between parallel branches**

This table compares what happens in the water circuit with what happens in the electric circuit.

Water circuit		Electric circuit
the battery		the pump
three water wheels in series		three resistors in series
the water flows round the circuit		the charges flow round the circuit
the pump does work on the water to raise it to a higher level	is like	the battery does work on the electric charges to 'lift' them up to a 'higher energy level'
the water does work on the water wheels and energy is transferred		the charges do work on the resistors and energy is transferred
the increase in gravitational potential energy when the water is pumped to the high level is the same as the decrease when the water drops to the low level		the potential differences across the resistors add up to the potential difference across the battery

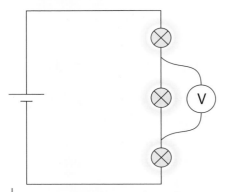

There is a reading on this voltmeter. it is measuring the potential difference between two points.

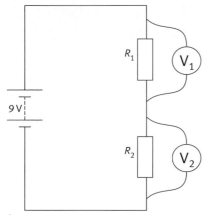

The voltages add up to the battery voltage.

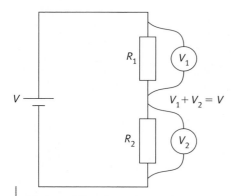

Two resistors in series make a potential divider.

A voltmeter measures the difference in 'level' between the two points it is connected to. This is called the potential difference between these points. **Potential difference (p.d.)** is measured in volts (V).

The important things to remember about potential difference in series circuits are:
- The voltage of a battery is the potential difference between its terminals.
- The sum of the p.d.s across them is equal to the battery voltage.
- The p.d. across each resistor depends on its resistance. The biggest voltmeter reading is across the resistor with biggest resistance.

Potential divider

This is a circuit with two resistors in series. It has many important uses.
- The sum of the potential differences across the two resistors is equal to the p.d. across the battery (the battery voltage).
- The resistors divide up the battery voltage into two parts.
- The bigger p.d. is always across the bigger resistance.
- If the resistors are equal the p.d.s across them are equal.

There is a potential difference of 12 V across the terminals of a car battery.

Question

1 In the circuit above, the resistors R_1 and R_2 are the same.
   a What are the readings on V_1 and V_2?
   b R_2 is replaced with a bigger resistance. Which has a higher reading, V_1 or V_2?

Currents in parallel circuits

When different resistors are connected in parallel, the largest current is through the smallest resistor.

This diagram shows water flowing through a large pipe. The pipe splits in two and then joins up again. When the two parallel pipes have different diameters, more water flows every second through the pipe with the larger diameter. The wider pipe has less resistance than the narrower pipe to the flow of water. So the current through it is larger.

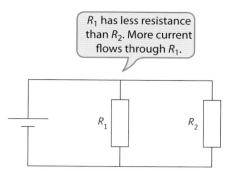

R_1 has less resistance than R_2. More current flows through R_1.

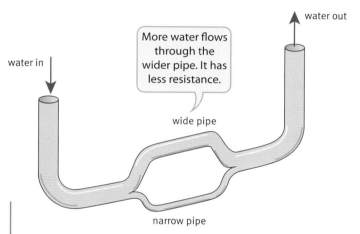

More water flows each second through the larger pipe. It has less resistance to the water flow.

When different resistors are connected in parallel the total current from the battery is the sum of the currents through the resistors.

This diagram shows two resistors connected in parallel to a battery as making two separate loop circuits that share the same battery. One is shown in red and the other in purple. Some wires in the circuit are part of both loops. This is where the current is biggest. The current here is the sum of the currents in the loops.

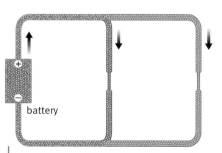
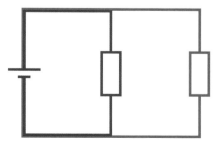

A parallel circuit behaves like two separate loop circuits.

Summary box

- Potential difference (p.d.) is another way of saying voltage.
- A voltmeter is used to measure the p.d. between two points.
- When two components are connected in series to a battery:
  - the p.d. across the components adds up to the p.d. across the battery
  - the p.d. is largest across the component with the largest resistance.
- When components are connected in parallel to a battery:
  - the current is largest through the smallest resistance
  - the total current to (and from) the battery is the sum of the currents through the branches.

F Electrical power

Find out about

✓ how the power produced in a circuit component depends on both current and voltage

An electric circuit transfers energy stored in the battery to somewhere else. A key feature of any electric circuit is the rate at which energy is transferred from the battery to the other components and on into the environment. This is called the **power** of the circuit.

Measuring the power of an electric circuit

In general, the power used in an electric circuit is:

power	=	current	×	voltage
P	=	I		V
(watt, W)		(ampere, A)		(volt, V)

The unit of power is the watt (W). One watt is equal to one joule per second.

You can show this is true with a battery and bulb circuit. There are two ways to double the power. The diagram shows how.

The rate of energy transfer is how much energy is transferred each second.

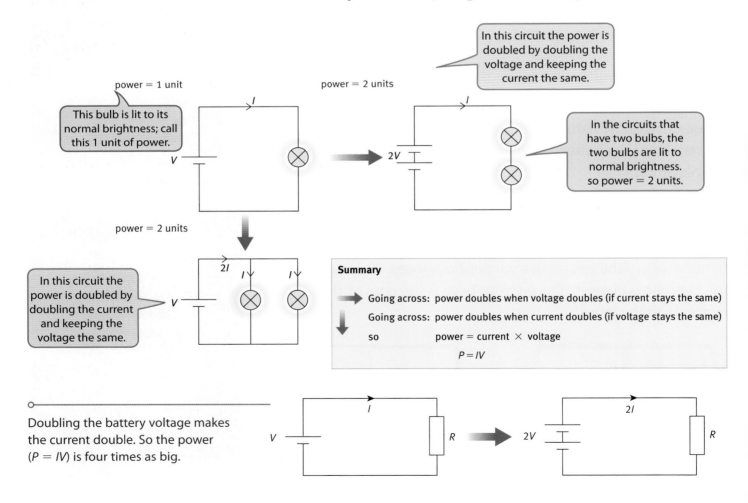

power = 1 unit

This bulb is lit to its normal brightness; call this 1 unit of power.

power = 2 units

In this circuit the power is doubled by doubling the voltage and keeping the current the same.

In the circuits that have two bulbs, the two bulbs are lit to normal brightness. so power = 2 units.

power = 2 units

In this circuit the power is doubled by doubling the current and keeping the voltage the same.

Summary

→ Going across: power doubles when voltage doubles (if current stays the same)

↓ Going across: power doubles when current doubles (if voltage stays the same)

so   power = current × voltage

$P = IV$

Doubling the battery voltage makes the current double. So the power ($P = IV$) is four times as big.

If you know the power, it is easy to calculate how much work is done (or how much energy is transferred) in a given period of time:

work done (or energy transferred) = power × time
(joules, J) (watts, W) (seconds, s)

Worked example

A 12 V motor is connected to a battery and is used to lift a load. The current is 8 A. What is the power of the motor?

power = current × voltage
= 8 A × 12 V
= 96 W

The load is lifted in 20 seconds. How much work has the motor done?

work done = power × time
= 96 W × 20 s
= 192 J

The power of the electric motor in this tube train is much greater than the power of the strip light above the platform. Both the voltage and the current are bigger.

Questions

1. A 9 V battery is connected to a lamp and the current is 0.3 A.
   a What is the power of the lamp?
   b How much energy is transferred in 10s?

2. In circuit A, a battery is connected to a resistor with a small resistance. In circuit B, the resistor has a large resistance. The two batteries are identical. Which will go 'flat' first? Explain your answer.

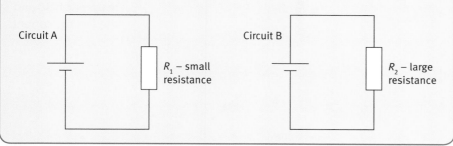

Summary box

- Power = energy transferred each second.
- Power = current × voltage.
- Energy transferred = power × time
- Power is measured in watts.

G | Magnets and motors

Find out about

- the force on a current-carrying wire in a magnetic field
- why an electric motor spins

Magnetic effects

When an electric current is switched on and off in a wire, the compass needle of a nearby magnetic compass moves. This is because it is in a magnetic field caused by the electric current.

Winding a wire into a coil makes the magnetic field around it stronger. This is because the fields of each turn of the coil add together. It can be strengthened further by putting an iron core inside the coil to make an **electromagnet**.

Magnetic forces

A permanent magnet experiences a force if it is placed in the magnetic field near a wire that is carrying an electric current. In the diagram the magnet is fixed and the electric current is carried by a piece of wire. When the current is switched on:

- the wire 'rider' moves sideways
- the movement is always at right angles to the magnetic-field direction and the current direction
- there is no effect if the wire is parallel to the magnetic field.

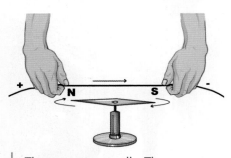

The compass needle. The needle moves when the electric current is switched on.

A magnet has a north pole, N, and a south pole, S. The magnetic-field direction is from one pole to the other.

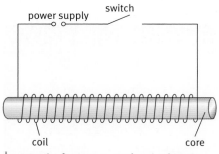

A coil of wire wound round an iron core makes an electromagnet – a magnet that can be switched on and off.

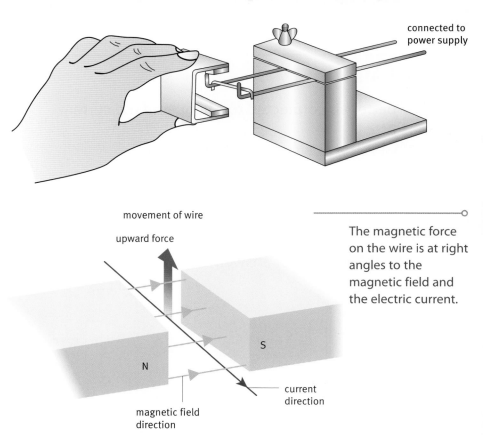

The magnetic force on the wire is at right angles to the magnetic field and the electric current.

Turning effect on a coil of wire

Magnetic forces can make a wire coil turn when a current flows.

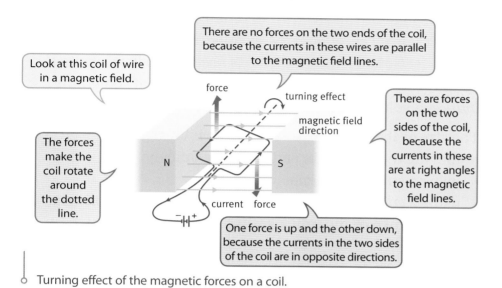

Turning effect of the magnetic forces on a coil.

When you use a coil with lots of turns the forces are stronger.

An electric motor

To use this effect in an electric motor we have to keep the coil turning. To do this the coil is connected to the battery using a commutator.

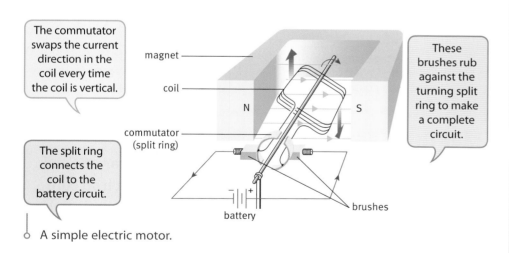

A simple electric motor.

This flowchart shows how the coil keeps rotating.

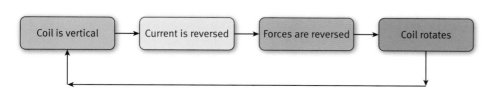

This computer has three motors – one for the hard disc drive, one for the DVD drive, and one to turn the cooling fan.

Summary box

- A current at right angles to a magnetic field produces a force.
- There is no force if the current is parallel to the magnetic field.
- The force produces movement, for example, a coil can be turned.
- To keep the coil turning a commutator reverses the current direction every half turn.

Question

1. Look at the diagram. What would happen if:
   a the magnet N and S were swapped?
   b the current was reversed?
   c both a and b were done together?

H Generating electricity

Find out about

- how a magnet moving near a coil can generate an electric current
- the factors that affect the size of this current
- how this is used to generate electricity on the large scale

Electromagnetic induction

You can generate an electric current by moving a magnet into a coil of wire. This effect is called **electromagnetic induction**. When electromagnetic *induction* is used to produce a voltage, you say the voltage is an *induced* voltage, or that something *induces* a voltage.

The size of the induced voltage can be increased by:
- moving the magnet in and out more quickly
- using a stronger magnet
- using a coil with more turns so that the induced voltages in each turn of the coil, together.

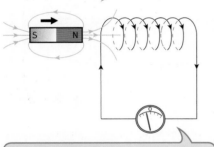

1. While the bar magnet is moving into the coil, there is a small reading on the sensitive ammeter.

The magnetic field around the wires of the coil changes. Magnetic field lines are 'cutting' the coil. This induces a voltage across the coil. If the coil is connected into a complete circuit, this voltage causes a current.

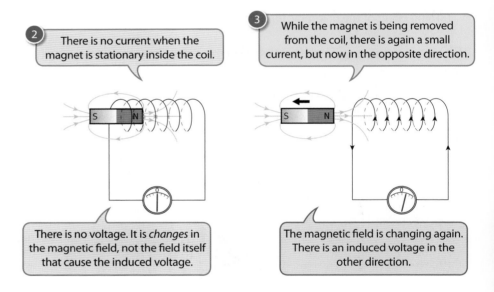

2. There is no current when the magnet is stationary inside the coil.

There is no voltage. It is *changes* in the magnetic field, not the field itself that cause the induced voltage.

3. While the magnet is being removed from the coil, there is again a small current, but now in the opposite direction.

The magnetic field is changing again. There is an induced voltage in the other direction.

Making a generator

We can make a **generator** by rotating a magnet near one end of a coil. The induced voltage is bigger if we put an iron core into the coil. This increases the strength of the magnetic field inside the coil. As the magnet rotates, the magnetic field around the coil is constantly changing.

This induces a voltage across the ends of the coil, which causes an electric current in the circuit. The voltage and current change direction every half-turn of the magnet. This results in an **alternating current (a.c.)** in the circuit. An alternating current is one that keeps changing direction.

For many applications, a.c. works just as well as the **direct current (d.c.)** produced by a battery. A direct current is one that flows in just one direction. Heaters and filament lamps, for example, make use of the heating effect of a current. It doesn't matter in which direction the current flows.

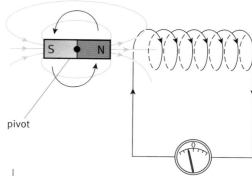

When the magnet rotates, it induces an alternating voltage across the coil.

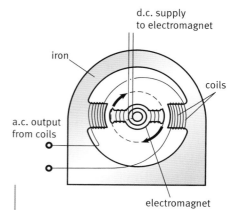

An a.c. generator. An electromagnet is rotated in a coil of wire.

Generators in a power station.

Questions

1 On these pages find four ways of increasing the voltage from a generator.

2 Describe the difference between alternating current produced by a power station and the direct current from a battery.

Summary box

- When a magnet moves in a coil of wire a voltage is induced.
- Mains electricity is produced by generators using electromagnetic induction.
- Batteries produce d.c., mains is a.c.

I Distributing electricity

Transformers

Transformers are very useful. You probably have chargers for mobile phones and other electronic equipment. These contain transformers and so do a lot of electronics, like TVs and PCs.

Transformers change the voltage. Step-down transformers change it to a lower value and step-up transformers change it to a higher value.

The diagram shows how a transformer is constructed. It has two completely separate coils wound on an iron core. There is no current flow between the coils. They are in separate circuits.

Find out about

- how transformers are used to alter the voltage of a supply
- the main components of the National Grid

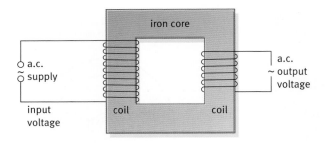

The coils are linked by the magnetic field in the iron core. When an alternating current flows in one coil an alternating voltage is induced in the other coil. It is very important that the supply is a.c., otherwise the magnetic field in the core will not be changing and no voltage is induced. Transformers only work with a.c.

Mains electricity

In a typical power-station generator an electromagnet is rotated inside a fixed coil. As it spins a.c. is generated in the coil.

Questions

1. A phone uses 4.5 V and the charger plugs into 230 V mains. What type of transformer is in the charger?
2. A 5000 V d.c. power supply is available, but a 1000 V d.c. supply is needed. Explain whether a transformer could be used to step down the voltage.

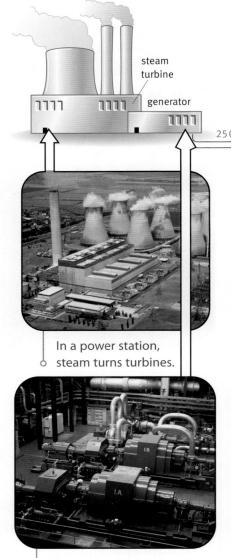

In a power station, steam turns turbines.

The turbines turn the generators.

P5: ELECTRIC CIRCUITS

The National Grid

Transformers play an important role in the National Grid system. The National Grid distributes electricity from the power stations to the rest of the country.

Transformers are used to step up the voltage from the power stations and then step down again at the local substations. The mains voltage supplied to our homes in the UK is 230 V.

> **Summary box**
> ✓ Transformers:
> - can be used to increase or decrease an alternating voltage
> - are made of two separate coils on the same iron core
> - only work with an a.c. supply.
> ✓ Mains supply in the UK is 230 V a.c.

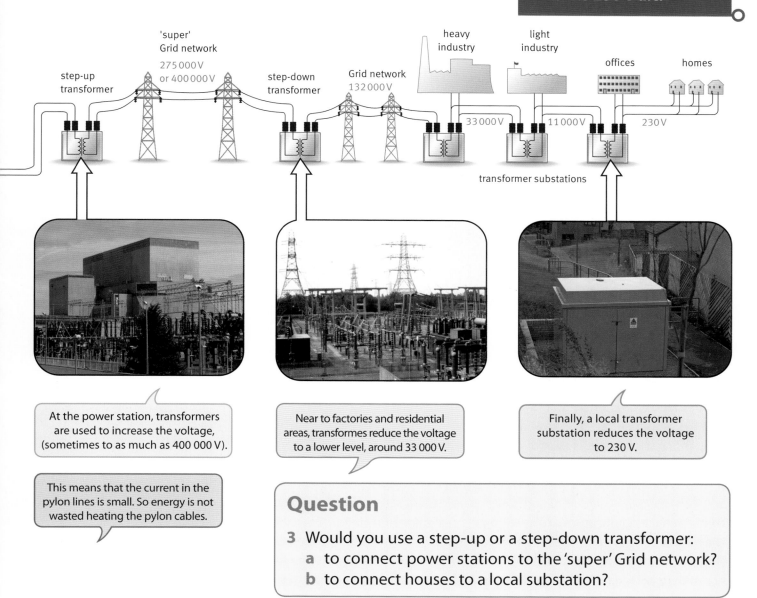

At the power station, transformers are used to increase the voltage, (sometimes to as much as 400 000 V).

This means that the current in the pylon lines is small. So energy is not wasted heating the pylon cables.

Near to factories and residential areas, transformes reduce the voltage to a lower level, around 33 000 V.

Finally, a local transformer substation reduces the voltage to 230 V.

Question

3 Would you use a step-up or a step-down transformer:
   a to connect power stations to the 'super' Grid network?
   b to connect houses to a local substation?

I: DISTRIBUTING ELECTRICITY

Science Explanations

Electricity is essential to modern-day life. An understanding of electric charge, current, voltage, and resistance in a circuit allows us to use electricity safely and enables power to be generated and distributed.

You should know:

- about electric charge and how positive and negative charges can be separated
- that electric current is a flow of charges already present in the materials of the circuit
- how electric circuits work, and about models that help us understand electric circuits
- that current is not used up as it goes around but it transfers energy from the battery to other components
- that the voltage of a battery is a measure of the push on the charges
- that the bigger the voltage the bigger the current
- that the components in a circuit resist the flow of charge and how the current depends on the battery voltage and the circuit resistance
- that resistors get hotter when current flows through them and that is why filament lamps glow
- about components with a variable resistance, including thermistors and light-dependent resistors (LDRs)
- how to measure the voltage between two points using a voltmeter
- that the voltage is also called the potential difference (p.d.)
- about the p.d. across and the current through resistors connected in series and in parallel
- about the power (energy per second) transferred by an electric circuit
- about the force on a current-carrying wire in a magnetic field and why a motor spins
- about electromagnetic induction, including:
  - a p.d. is induced across the ends of a wire, or coil, in a changing magnetic field
  - if this wire, or coil, is part of a circuit there is an induced electric current in the circuit
  - the magnetic field must be changing, otherwise there is no effect
  - how the effect is increased and how it is used in making an electrical generator
- that electrical generators are used to produce mains electricity
- the difference between a.c. and d.c. electricity
- that the electricity supply to our homes is 230 V a.c.
- about transformers and the effect of changing the number of turns in the coils.

components in series

generators

transformers

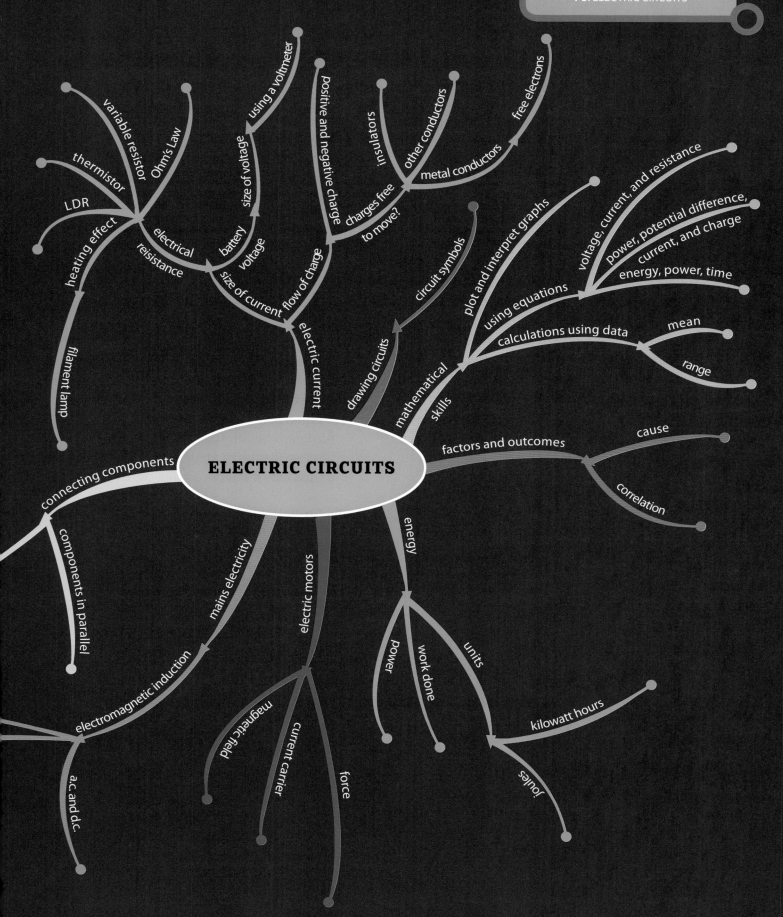

Ideas about Science

In addition to developing an understanding of electric circuits, it is important to understand how scientists use data to develop their ideas.

Collecting data is often the starting point for a scientific enquiry, but data can never be trusted completely. Data is more reliable if it can be repeated; when making several measurements of the same quantity, the results are likely to vary. This may be because:
- you have measured several individual samples, for example, several samples of a resistance wire
- the quantity you are measuring is varying, for example, the light level in the room is varying as you measure the resistance of an LDR
- there are limitations in the measuring equipment, for example, a poor electrical connection in the circuit.

Usually the best estimate of the true value of a quantity is the mean of several repeated measurements. When there is a spread of values in a set of measurements, the true value is probably in the range between the highest and the lowest values.

You should:
- be able to calculate the mean from a set of repeat measurements
- know that a measurement may be an outlier if it is well outside the range of other measurements.

When comparing sets of data to decide if there is a difference between the two means, it is useful to look at the ranges of the data. You should know that:
- if the ranges of two sets of data do not overlap there is probably a real difference between the means.

To investigate the relationship between a factor and an outcome, it is important to control all the other factors that might affect the outcome. In a plan for an investigation you should be able to:
- recognise that the control of other factors is a positive feature of an investigation and it is a design flaw if factors are not controlled.

Factors and outcomes may be linked in different ways and it is important to distinguish between them. A correlation between a factor and an outcome does not necessarily mean that the factor causes the outcome; both might be caused by some other factor. For example, the more electricity substations there are in an area, the more babies are born in that area. But this is because there are more houses needing an electricity supply where more people live. You should be able to:
- identify a correlation from data, a graph, or a description
- explain why an observed correlation does not necessarily mean that the factor causes the outcome
- explain why individual cases do not provide convincing evidence for or against a correlation.

Review Questions

P5: ELECTRIC CIRCUITS

1. Look at the electric circuit models in this module. Copy and complete the following table.

Model	What corresponds to: the battery?	electric current?	the resistors or lamps?
'peas in a pipe'			
'water in a pipe'			

2. Imagine a simple electric circuit consisting of a battery and a bulb. For each of the following statements, say if it is true or false (and explain why):
   a Before the battery is connected, there are no electric charges in the wire. When the circuit is switched on, electric charges flow out of the battery into the wire.
   b Collisions between the moving charges and fixed atoms in the bulb filament make it heat up and light.
   c Electric charges are used up in the bulb to make it light.

3. You are given four 4 Ω resistors. Draw diagrams to show how you could connect all four together to make a resistance of:
   a 16 Ω   b 1 Ω

4. Anna is given a resistor but it is not labelled with the value of the resistance.
   a Write down the equation that she could use to calculate the resistance.
   b What does she need to measure to work out the resistance?
   c Draw a circuit diagram to show how she could measure the quantities in your answer to part b.

5. Peter has a sensor labelled LDR.
   a What do the letters LDR stand for?
   b What does an LDR detect?
   c Suggest what Peter might use an LDR for in a sensing circuit.

6. Sam has a thermistor.
   a What does a thermistor detect?
   b Suggest what Sam might use a thermistor for in a sensing circuit.

7. Copy and complete these sentences:
   When a magnet is moved into a coil of wire, a voltage is _____ in the coil. The voltage is produced only when the magnet is _____. This is used in an a.c. generator, which has an _____ rotating near a fixed coil. To increase the size of the induced voltage, you could use a _____ electromagnet, have more _____ on the fixed coil, turn the rotor coil _____, or put a core of _____ inside it. The current in the external circuit constantly changes direction, so it is called _____ current (__). This is different from the current from a battery, which always goes in one direction and is called current (__).

8. What are the similarities and differences between a motor and a generator?

9. A school laboratory has a set of transformers to demonstrate how power lines work. The transformer has 240 turns on the primary coil and 1200 turns on the secondary coil. How will the output voltage be different to the input voltage?

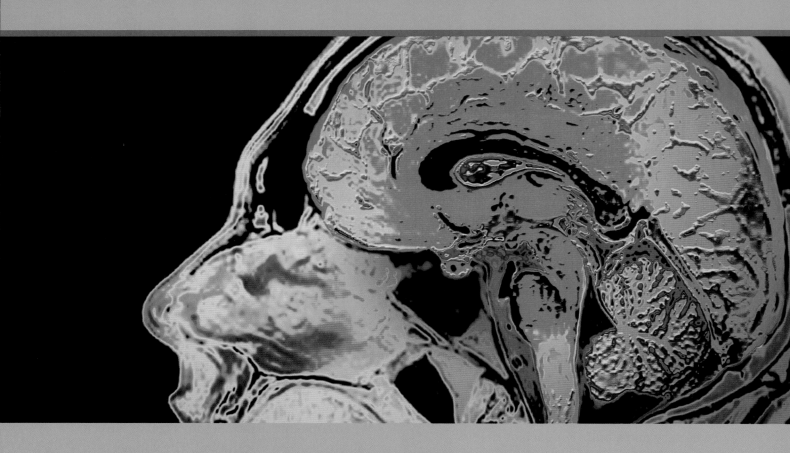

B6 Brain and mind

Why study the brain and mind?

The human brain allows our species to survive on Earth. It gives us advantages of intelligence and sophisticated behaviour.

What you already know

- The success of humans is mainly due to the evolution of a large and complex brain.
- Drugs can affect human behaviour.
- Nerves and hormones help you respond to your environment.

Find out about

- how living things respond to stimuli
- how nerve impulses are passed around your body
- how your brain processes information from your senses
- how you learn new skills
- how scientists are finding out about memory.

The Science

Animals respond to stimuli in order to survive. The brain and spinal cord coordinate millions of electrical impulses every second. These impulses shape how we think, feel, and react – our behaviour. Some drugs can affect this. The structure of human brains allows us to learn and recall large amounts of information.

Ideas about Science

Scientific research has ethical implications. Some people say that the right decision is the one leading to the best outcome for the most people. Others say that some actions are always wrong. This must be considered when researching the brain.

A What is behaviour?

Find out about

- what behaviour is
- how simple behaviour helps animals survive

Imagine you are sitting outside. The temperature drops, and you get cold. You start to shiver.

Shivering is a **response** to the change in temperature. A change in your environment is called a **stimulus**. Eating is a response to the stimulus of hunger. Scratching is a response to an itch. Shivering, eating, and scratching are all examples of **behaviour**.

You can think of behaviour as anything an animal does. The way an animal responds to changes in its surroundings is important for its survival.

Simple reflexes help animals to survive

Simple animals always respond to a stimulus in the same way. For example, woodlice always move away from light. This is an example of a **simple reflex** response. Reflexes are always **involuntary** – they are automatic. The photographs in this section all show reflexes.

Simple reflex behaviour helps an animal to:
- find food, shelter, or a mate
- escape from their enemies – predators
- avoid harmful places, for example, high temperatures.

However, animals with only simple reflexes cannot change their behaviour, or learn from experience.

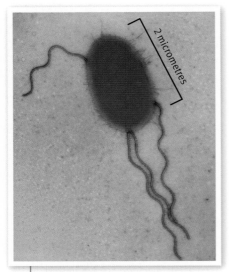

Escherichia coli bacteria are found in your gut. They detect the highest concentration of food and move towards it.

Woodlice move away from light, so you are more likely to find them in dark places.

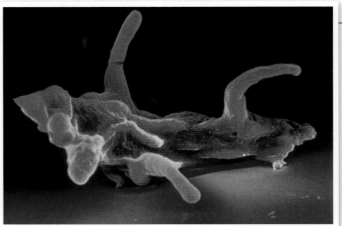

Single-celled *Amoeba* live in pond water. They move away from high concentrations of salt, strong acids, and alkalis.

When a giant octopus sees an enemy, it quickly contracts its body muscles. This squirts out a jet of water, which pushes the octopus away from danger. The octopus may also release a dark chemical (often called 'ink'), which hides its escape.

When the tail of the sea hare *Aplysia* is pinched, the muscle contracts quickly and strongly. This reflex helps the animal escape from the spiny lobster that preys on it.

Earthworms have some of the fastest reflexes in the animal kingdom. A sharp tap from a beak on its head end is detected in the body wall. The worm's muscles pull it back quickly into its burrow. But this time the bird was too fast.

Have you ever tried to swat a fly? A fly's eyes are very sensitive to movement. It can respond very quickly.

Summary box
- The way that animals respond to the different stimuli around them is known as behaviour.
- Simple reflexes are automatic responses that help animals survive.
- Complex behaviour in higher animals involves conscious decisions and learning.
- Complex behaviour gives greater chances of survival.

Questions
1. Describe an action you did today that:
   - you did not have to think about and you have never had to learn to do
   - you have learnt to do but you can now do without thinking
   - you had to think about while you were doing it.
2. Which of the actions you described is an example of complex conscious behaviour? Which is most likely to be a simple reflex response?

Complex behaviour – a better chance of survival

Complex animals, like mammals, birds, and fish, have simple reflex responses too. But a lot of their behaviour is far more complicated. Much of their behaviour involves making conscious decisions. For example, if it gets very cold, you do not just rely on your reflexes to keep you warm; you decide to put on extra clothes.

Because complex animals can change their behaviour when environmental conditions change, they are more likely to survive.

B | Simple reflexes in humans

Find out about
- reflexes in newborn babies
- simple reflexes that help you to survive

Behaviour in humans and other mammals is usually very complex. But simple reflexes are still important for survival. For example:
- When an object touches the back of your throat, you gag to avoid swallowing it. This is the gag reflex.
- When a bright light shines in your eye, your pupil becomes smaller. This **pupil reflex** stops bright light from damaging the sensitive cells at the back of your eye.
- When you pick up a very hot object you may quickly drop it, even before you have realised that it is hot.

These types of behaviour are inherited through our genes. This is called **innate** behaviour.

Newborn reflexes

When a baby is born, the nurse checks for a set of **newborn reflexes**. Many of these reflexes are only present for a short time after birth. They are gradually replaced by behaviours learned from experience.

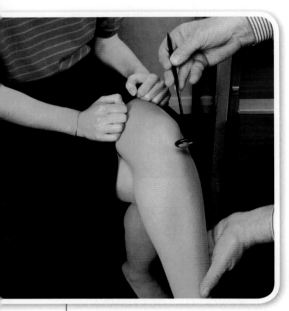

The knee jerk reflex makes your thigh muscle contract, so your leg straightens. Doctors may test this reflex when you have a health check. Try standing still with your eyes closed. You will notice this reflex helping you to balance.

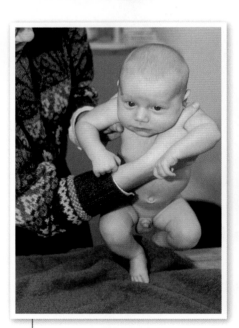

Stepping. If you hold a baby under his arms, support his head, and allow his feet to touch a flat surface, he will appear to take steps and walk.

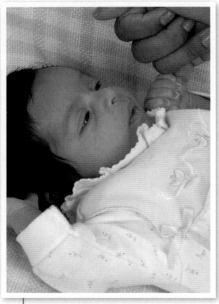

Grasping. When you put your finger in a baby's open palm, the baby grips the finger. When you pull away, the grip gets stronger. If you stroke the underneath of a baby's foot, its toes and foot will curl.

B6: BRAIN AND MIND

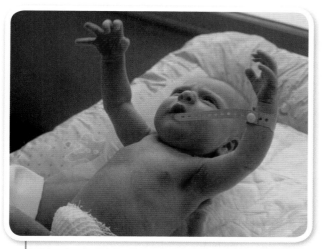

Startle. This reflex usually happens when a baby hears a loud noise or is moved quickly. The response includes spreading the arms and legs out and extending the neck. The baby then quickly brings her arms back together and cries.

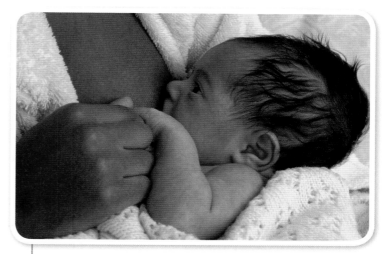

Sucking. Placing a nipple (or a finger) in a baby's mouth causes the sucking reflex. It is slowly replaced by voluntary sucking at around two months.

Rooting. Stroking a baby's cheek makes her turn towards you, looking for food. This reflex helps the baby find the nipple when she is breast feeding.

Swimming. If you put a baby under six months of age in water, he moves his arms and legs while holding his breath.

Summary box
- Humans have simple reflexes. They:
  - protect us from harm
  - are innate – inherited by everyone
  - include the gag reflex and pupil reflex.
- Newborn reflexes
  - are present at birth
  - include stepping, grasping, and sucking.

Questions

1. Describe two reflexes in:
   a. adult humans
   b. newborn babies.
2. How do you think the startle reflex helps a baby to survive?

B: SIMPLE REFLEXES IN HUMANS

Receptors detect stimuli

You can only respond to something if you can detect it. **Receptors** inside and outside your body detect stimuli, or changes in the environment.

You can detect many different stimuli, for example, sound, texture, smell, temperature, and light. Different types of receptors each detect a different type of stimulus. Receptors on the outside of your body monitor the external environment. Others monitor changes inside your body, for example, core temperature and blood sugar levels.

The eye is a complex sense organ

Some receptors are made up of single cells, for example, pain receptors in your skin. Other receptor cells are grouped together as part of a complex sense organ, for example, your eye.

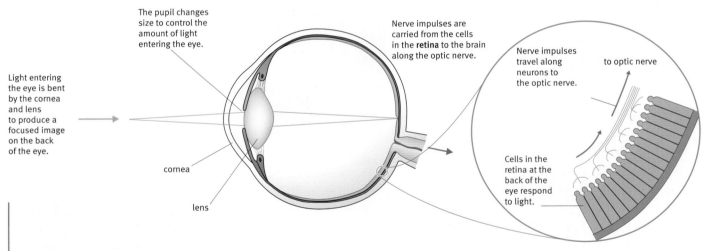

Light entering the eye is bent by the cornea and lens to produce a focused image on the back of the eye.

The pupil changes size to control the amount of light entering the eye.

Nerve impulses are carried from the cells in the **retina** to the brain along the optic nerve.

Nerve impulses travel along neurons to the optic nerve.

to optic nerve

Cells in the retina at the back of the eye respond to light.

cornea

lens

Light is focused by the cornea and lens onto light-sensitive cells at the back of the eye. These cells are receptors. They trigger nerve impulses to the brain.

Question

3 Some people suffer from a disease where tiny clusters of light receptor cells in different parts of the eye become damaged. How would this affect what the person sees?

Effectors are organs that respond

The body's responses to stimuli are carried out by **effector** organs. Effectors are either **glands** or **muscles**.

Summary box
- Receptors detect stimuli outside and inside our body.
- Some receptors are highly complex - like the eye.
- Effectors like glands and muscles let us respond to a stimulus.

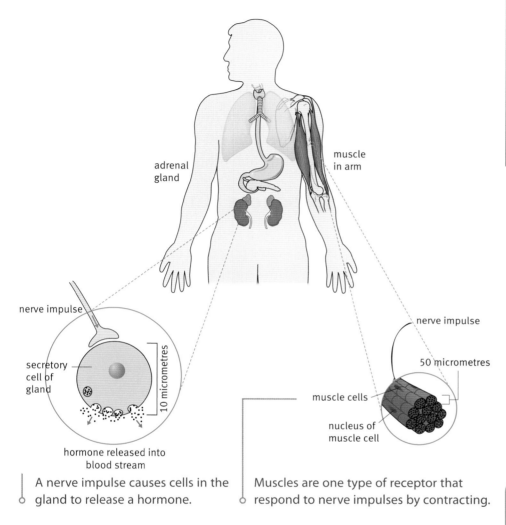

A nerve impulse causes cells in the gland to release a hormone.

Muscles are one type of receptor that respond to nerve impulses by contracting.

Nerve impulses bring about a fast but short-lived response in muscles. The muscles contract to move parts of the body. **Hormones** bring about longer-lasting effects such as an increase in growth rate.

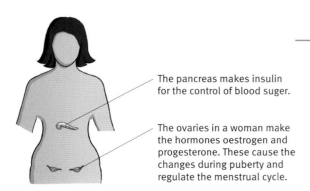

The pancreas makes insulin for the control of blood suger.

The ovaries in a woman make the hormones oestrogen and progesterone. These cause the changes during puberty and regulate the menstrual cycle.

Hormones are released when a nerve impulse reaches glands such as the ovaries and pancreas. Hormones have a slower and longer-lasting effect than a nerve impulse.

Questions

4. Name the two different types of effectors and say what they do.
5. Which receptors would you use in order to thread your trainers with new laces?
6. Which effectors are you using when you:
   - text a friend?
   - cry?
   - run a race?

C — Your nervous system

Find out about

- different parts of your nervous system
- how reflexes are controlled

Muscles in the iris cause the pupil to change size. A bright light stimulus triggers the pupil to shrink (top).

Evolution has produced complex multicellular animals. These animals developed nervous systems to allow them to respond to the environment. This gave them a survival advantage over simpler animals.

The **nervous system** is made up of **neurons** linking receptor cells to effector cells. It coordinates the body's responses.

Cells in your nervous system carry **nerve impulses**. These electrical impulses allow the different parts of the nervous system to communicate with each other.

The reflex arc produces automatic responses

In a simple reflex, impulses are passed from one part of the nervous system to the next in a pathway called a **reflex arc**. The diagram below shows this pathway for a pain reflex.

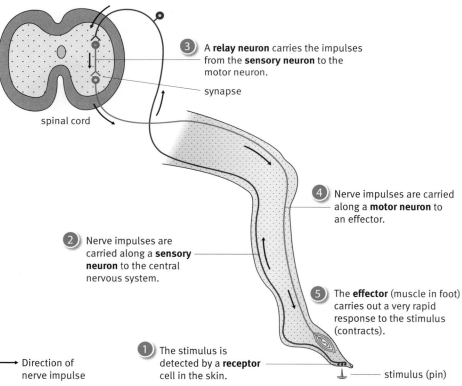

1. The stimulus is detected by a **receptor** cell in the skin.
2. Nerve impulses are carried along a **sensory neuron** to the central nervous system.
3. A **relay neuron** carries the impulses from the **sensory neuron** to the motor neuron.
4. Nerve impulses are carried along a **motor neuron** to an effector.
5. The **effector** (muscle in foot) carries out a very rapid response to the stimulus (contracts).

→ Direction of nerve impulse

A simple reflex arc. Your body responds to the pain stimulus in the correct way – even before you have realised what has happened.

Summary box

- Nerves are made up of specialised, long nerve cells called neurons.
- The central nervous system (CNS) is made up of the brain and spinal cord.
- The peripheral nervous system links the CNS to every part of your body.
- The reflex arc is automatic and fast.

Nerves and nerve cells (neurons)

Nerves are bundles of specialised cells called neurons. Like most body cells, neurons have a nucleus, a cell membrane, and cytoplasm. They are different from other cells because the cytoplasm is very long and thin in shape. This is called the **axon**, and it is how neurons connect different parts of the body.

Axons carry electrical nerve impulses – like wiring in an electrical circuit. The axons must be insulated from each other. The insulation for an axon is a **fatty sheath** wrapped around the outside of the cell. The fatty sheath also increases the speed that impulses move along the axon.

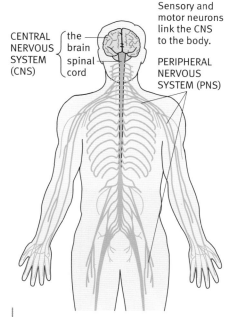

The central nervous system is made up of the brain and spinal cord.
The peripheral nervous system links the brain and spinal cord with the rest of the body.

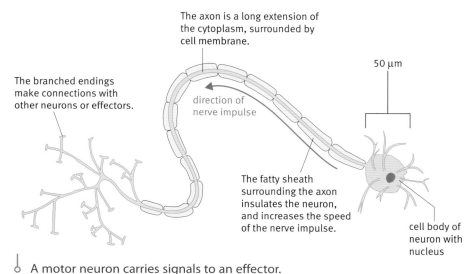

A motor neuron carries signals to an effector.

Central nervous system

Your **central nervous system** (CNS) coordinates all the information it receives from your receptors. Information about a stimulus goes to your spinal cord and often your brain. In a reflex arc the CNS directly links the incoming information from receptors with the effectors that will carry out the necessary response.

Peripheral nervous system

Many nerves link your brain and spinal cord to every other part of your body. These nerves make up the **peripheral nervous system**.

Questions

1. Describe the difference between:
   - the job of a sensory neuron and a motor neuron
   - the central nervous system and the peripheral nervous system.

2. Draw a labelled diagram to show a reflex arc for the newborn grasp reflex shown in Section B. This reflex is coordinated by the spinal cord.

D | Synapses

Find out about

✓ how nerve impulses pass from one neuron to the next

Nerve impulses give you fast reactions because they travel along axons at 400 metres per second.

A synapse is a gap between neurons

Neurons do not touch each other. So when nerve impulses pass from one neuron to the next, they have to cross tiny gaps. These gaps are called **synapses**. Some drugs and poisons (toxins) interfere with nerve impulses crossing a synapse. This is how they affect the human body.

Nerve impulses cannot jump across the gap between neurons at a synapse. Instead, chemicals are used to pass an impulse from one neuron to the next.

Quick responses to stimuli are important in fast-moving sports – but they also help you survive by avoiding danger.

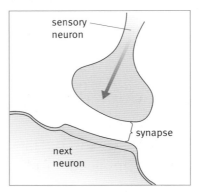

1. A nerve impulse arrives at a synapse. The direction of the impulse is shown by the arrow.

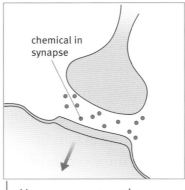

2. A chemical is released into the synapse. This diffuses across to the next neuron. A nerve impulse is stimulated in the next neuron.

How a synapse works.

Synapses help to coordinate nerve impulses

The impulse can only travel in one direction across the synapse. This makes sure that impulses travel in the right direction. The gap at the synapse is only about 20 nanometres (nm) wide. It only takes a fraction of a second for a nerve impulse to cross a synapse.

Being human – just chemicals in your brain?

Chemicals that interfere with the signals at the synapses in your brain have powerful effects. Some chemicals change the way we feel, alter what we see and hear around us, or even cause paralysis and death.

The drug ecstasy interferes with synapses

Ecstasy is the common name for the drug MDMA. It affects the way nerve impulses cross synapses in the brain. Ecstasy gives people feelings of happiness. It makes them feel very close to other people. Studies on monkeys suggest that long-term use of Ecstasy may destroy the synapses in the pleasure pathways of the brain. Permanent anxiety and depression might result, along with poor attention span and memory. Ecstasy can also be harmful because it interferes with the body's temperature and water balance.

Prozac is used to treat depression

Depression can be caused when the amount of synapse transmitter substance in the brain is too low. **Prozac** works by stopping the transmitter substance from being 'mopped up' at the synapses. Like all drugs, Prozac can have unwanted effects.

Beta blockers treat heart pains

Some people get bad chest pains. Nerve impulses stimulate the heart to speed up when people are stressed or excited. This can leave the heart muscle painfully starved of oxygen. Beta blockers reduce the impulses crossing nerve synapses, which stops the heart from speeding up. Beta blockers also help to control nerve impulses inside the heart, making sure that the heart beats in a regular, controlled way.

Curare is a very powerful nerve toxin. It is used on the tips of blowpipe darts. Even a tiny amount of toxin on the dart is enough to paralyse or kill an animal. The toxin stops nerve impulses from crossing synapses.

Summary box
- At the junction between two neurons is a gap called a synapse.
- The nerve impulse crosses the synapse when a chemical is released.
- Chemicals that interfere with the signals at the synapse have powerful effects.

Questions

1 Write down a sentence to describe a synapse.
2 What is a nerve toxin?

E. The brain

Find out about
- the structure of your brain
- how scientists learn about the brain

Simple animals have simple brains

Receptors produce impulses. Impulses are sent to effectors. If an animal is to survive it must respond correctly to stimuli. Even very simple animals have brains so they can process information and coordinate their response.

Complex animals

More complex behaviour like yours needs a much larger brain. So your brain is made of billions of neurons. It also has many areas, each carrying out one or more specific functions all in the same organ. Your complex brain allows you to learn from experience, for example, how to behave with other people.

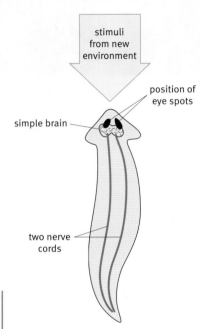

Sense organs on the flatworm head detect light and chemical stimuli. A simple brain processes the response.

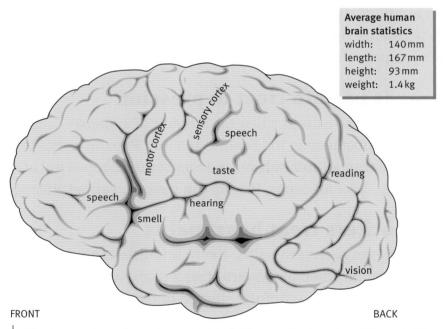

Average human brain statistics
width: 140 mm
length: 167 mm
height: 93 mm
weight: 1.4 kg

The human brain has a large, highly folded cerebral cortex. Although it is only 5 mm thick, its total area is about 0.5 m^2. This map of the cerebral cortex shows the regions responsible for some of its functions.

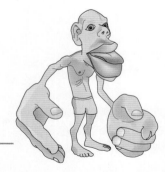

This is a diagram called the 'sensory homunculus'. Each body part is drawn so that its size represents the surface area of the brain's sensory cortex that receives nerve impulses from it.

The conscious mind

When you are awake, you are aware of yourself and your surroundings. This is called **consciousness**. The part of your brain where this happens is the **cerebral cortex**. This part is also responsible for intelligence, language, and memory.

The cerebral cortex is very large in humans compared with other mammals.

Finding out about the brain

Scientists who study the brain are called **neuroscientists**. In the 1940s a Canadian brain surgeon, Wilder Penfield, was working with patients who had epilepsy. Penfield applied electricity to the surface of their brains in order to find the problem areas. The patients were awake during the operations. There are no pain receptors in the brain so they did not feel pain.

Penfield watched for any movement the patient made as he stimulated different brain regions. From this information he was able to identify which muscles were controlled by specific regions of the motor cortex.

Brain injuries give clues about brain function

Scientists study patients whose brains have been damaged, for example, by a bullet or a blood clot (a stroke). This tells them which part of the brain does which job. A patient with damage to their motor cortex might not be able to move part of their body. This shows scientists which part of the body that area of the brain controls.

Brain scans show the brain at work

Modern magnetic resonance imaging (MRI) brain scanners provide detailed information about the brain without having to open up the skull. MRI can be used to show which parts of the brain are most active when a patient does different tasks. These scans are called functional MRI (fMRI) scans. The active parts of the brain use more oxygen.

Questions

1. What is your brain made up of?
2. Why is a complex brain so important for survival?
3. What functions of the brain happen in the cerebral cortex?
4. Explain why it is necessary for blood flow to increase to parts of the brain that are very active.
5. What ethical issues should scientists consider when using injured humans to study the brain?

Summary box

- The brain processes information from receptors and sends the correct response to effectors.
- The cerebral cortex is responsible for intelligence, memory, feelings, and consciousness.
- Scientists get clues about how the brain works by studying injured patients and brain scans.

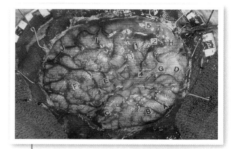

The brain surgeon Penfield mapped the motor cortex by stimulating the exposed human brain during surgery. Regions of this brain have been identified and labelled in a similar way.

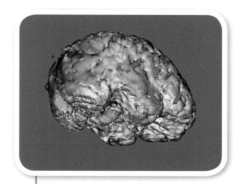

This functional MRI brain scan shows up areas of the brain that are active as a patient carries out a specific task. This patient was reading out loud.

F | Learnt behaviour

> **Find out about**
> - how conditioned reflexes can help you survive

The lion cub below is just a few weeks old. She was born with reflexes that are helping her to stay alive. She will **learn** much of her behaviour from her mother. For example, she will learn how to hunt or how to get on with other lions in the pride. This is called learnt behaviour.

Pavlov's experiment

Pavlov's dog salivated when presented with food.

The food is the **primary** stimulus and salivation is the response.

Pavlov rang a bell while his dog was eating its food.

After a while the dog salivated when it heard the bell, even if no food was around.

The dog had learnt to link the **secondary** stimulus of ringing the bell with food. This type of learning is called **conditioning**.

> Learning to link a new stimulus with a reflex action allows animals to change their behaviour. This is called a **conditioned reflex**.

Learnt behaviour is just as important for this lion cub's survival as reflexes.

Being able to learn new behaviour by experience is very important for survival. It means that animals can change their behaviour if their environment changes.

Reflex responses can be learnt

In 1904 the Russian scientist Ivan Pavlov won a Nobel Prize for his study on how the digestive system works. In his research Pavlov trained a dog to expect food whenever it heard a bell ring. This is called **conditioning**. The diagrams on the left explain what happened.

Conditioning your pet

Open a can of soup in your kitchen. If you have a dog or a cat, this sound may get them very interested. But they are not hoping for soup! The animal's reflex response to food has been conditioned. It has learnt through experience that the sound of a tin being opened may be followed by food being put into its dish.

If a cat only uses its basket when you are taking it to the vet for an injection, it may become conditioned to link the basket with a frightening experience. The cat will then always be frightened by the stimulus of the basket. It will fight to keep out of the basket, even if you are only trying to take it to a new home.

Goldfish become conditioned to expect food when they see you in the room. They swim to the front of the bowl when you appear.

Questions

1. Draw a flow diagram to explain how a cat or other pet can become conditioned to expect food when it hears a bathroom shower being run.
   Use the key words from this section in your answer.

2. Adverts often have funny or exciting images and catchy tunes. Write down a list of photos and tunes from adverts that remind you of things you could buy. How is conditioning involved in making us more likely to buy these products?

Summary box
- Animals can be conditioned to respond to a stimulus in a particular way.

Hundreds of neurons interact to coordinate the responses you make when you are receiving this many stimuli.

Our complex brain and complex behaviour

Connections in the brain usually involve hundreds of other neurons with different connections. Using these complex pathways, your brain can process highly complicated information, such as music, smells, and moving pictures. Different parts of the brain also store information (memory) and use it to make decisions for more complicated behaviour.

Complex behaviour allows us to learn from experience. For instance, **social behaviour** is learnt as humans develop.

Early humans learnt how to make and use tools for food and protection. Their ability to learn language meant that they could communicate new ideas. This gave them a survival advantage.

Hundreds of complex pathways in the brain have to be used to succeed in this fast-moving sport.

This complicated behaviour involves highly complex pathways in the brains of both these animals.

Experiments with living things involves ethical decisions

Our understanding of the brain and behaviour has mainly come from experimenting on animals and (in the past) humans. This has allowed us to improve our theories about human learning, and to develop new treatments for diseases and injuries.

Some people argue that using animals for medical research is alright. Other people think that experimenting with a vertebrate's brain for scientific reasons is wrong.

Scientists have learnt a lot about the brain from studying humans with mental-health disorders. But is it fair and right to experiment on patients that are ill? Some people argue that experimenting on ill patients results in improved medical knowledge that will benefit many others.

New technologies like magnetic resonance imaging (MRI) now mean that it is possible to see the brain working without causing any harm to the patient.

> **Summary box**
> ✓ Our brain is highly evolved, allowing us to process complex stimuli and respond correctly.
> ✓ Experimenting with living things involves ethical decisions.

Questions

3 Give three examples of how early humans' ability to learn gave them a survival advantage.

4 Do you think there is any ethical difference between using a rat and using a monkey for experiments to find out how the brain works?

5 Under what circumstances do you think it could be right to conduct scientific experiments on a human with a brain disorder?

G | Human learning

Find out about

- how human beings learn new things
- explanations that scientists have for how your memory works

When humans and other mammals experience something new, they can develop new ways of responding. Experience changes the way we behave, and this is called learning.

Learning involves building neuron pathways in the brain

Neurons in your brain are joined together to form complicated **pathways**. The first time a nerve impulse travels along a particular pathway, new connections are made between the neurons.

If the experience is repeated, or the stimulus is particularly strong, more nerve impulses follow the same nerve pathway. Each time this happens, the connections between these neurons are strengthened. Strengthened connections make it easier for nerve impulses to travel along a pathway. As a result, the response you produce becomes easier to make.

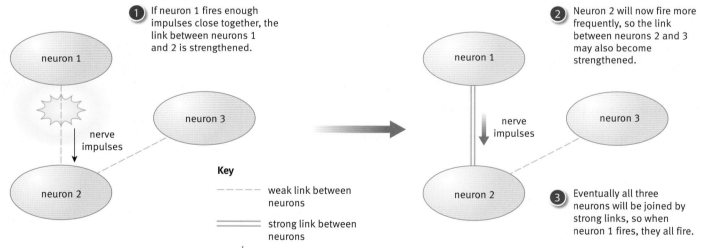

1. If neuron 1 fires enough impulses close together, the link between neurons 1 and 2 is strengthened.
2. Neuron 2 will now fire more frequently, so the link between neurons 2 and 3 may also become strengthened.
3. Eventually all three neurons will be joined by strong links, so when neuron 1 fires, they all fire.

Key
- - - - weak link between neurons
═══ strong link between neurons

Nerve pathways form in a baby's brain as a result of a stimulus from its environment. Repeating the stimulus strengthens the pathway. The baby then responds in the same way each time it receives the stimulus. Some neurons in the brain do not take part in any pathway. Many of these unused neuron pathways are lost.

The brains of human babies develop new nerve pathways very quickly. Your brain can develop new pathways all your life. This means you can still learn as you get older, though more slowly.

Repetition strengthens neuron pathways

Repetition helps you learn because it strengthens the pathways the brain uses to carry out a particular skill. Perhaps you have learned to ride a bicycle, play a musical instrument, perform a new dance sequence, or touch type. To do these things you created new neuron pathways then strengthened them through repetition. This made it easy for you to respond in the way that you practised.

Learning is easier when we are young

You learn to speak through repetition because you are surrounded by people talking. Children learn language extremely easily up to the age of about eight years. Their brains easily make new neuron pathways in the language processing region. As we get older it becomes harder for this part of the brain to make new pathways.

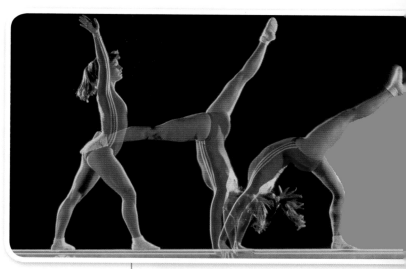

When Marie learns a new movement she imagines the position of her body and muscles being used at each stage. Thinking about using a muscle triggers nerve impulses to that muscle. This strengthens the pathways the impulse takes. Using visualisation, the actual movement is a lot easier to perform.

Questions

1. Write a few sentences to explain how you learn by experience Use the key words 'pathway' and 'repetition' in your answer.
2. Explain why repeating a skill helps you learn it.

Summary box

- Learning involves building neuron pathways in the brain.
- The more a pathway is stimulated the stronger the links become.
- We can learn by repeating or visualising an action.

H What is memory?

Find out about

- short-term and long-term memory
- the multistore model of memory
- the working-memory model

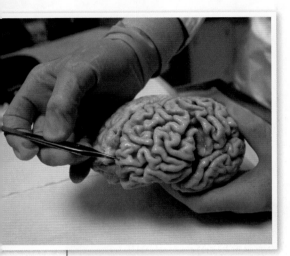

The 'Nun Study' at the University of Kentucky, USA, has had the participation of 678 School Sisters of Notre Dame. They ranged in age from 75 to 106 years. The nuns have allowed scientists to assess their health every year and to examine their brains at death. The study has helped scientists to understand brain disorders like Alzheimer's disease.

Psychologists are scientists who study the human mind. They describe **memory** as your ability to store and retrieve information.

Short-term memory

Read this sentence:

- As you read this sentence you are using your **short-term memory**.

Short-term memory lasts for about 30 seconds in most people. If you have no short-term memory you will not be able to make sense of this sentence. By the time you get to the end of the sentence, you will have forgotten the beginning.

Long-term memory

Think about a song you know the words to:

- To remember the words you use your **long-term memory**.

Long-term memory is a lasting store of information. There seems to be no limit to how much can be stored in long-term memory. And the stored information can last a lifetime.

Short-term memory and long-term memory work separately

People with advanced **Alzheimer's disease** suffer severe short-term memory loss. They cannot remember what day it is, or follow simple instructions. But they may still remember their childhood clearly.

Some people lose long-term memory because of brain damage or disease. Their short-term memory is normal. This evidence is important because it shows that long-term and short-term memory must work separately in the brain.

Sensory memory store

You use a sensory memory store to store sound and visual information for a short time. When you wave a sparkler on bonfire night it leaves a trail of light. You can even write shapes in the air that other people can see. You see the trail because you hold each image of the sparkler separately for a short time in your sensory memory store. The ability to hold images for a short time makes the separate pictures in a film seem continuous. You can store sound temporarily in the same way.

> **Summary box**
> - We have separate short-term and long-term memory stores in our brain.
> - Our sensory memory store helps us to understand actions and sounds.

Your sensory memory store holds each image of the sparkler separately for a short time. This makes the whole shape seem continuous.

Questions

1. Write down one sentence to describe memory.
2. What is the difference between short-term and long-term memory?
3. Explain why a person with advanced Alzheimer's disease is unable to do simple things like go shopping or cook for themselves.
4. Give one piece of evidence that short-term and long-term memory are separate.

How much can you hold in your short-term memory?

Cover up the list of letters below with a piece of paper. Move the paper down so you can see just the top row. Read through the row once, then cover it up and try to write down the letter sequence. Then go down to the next row and do the same. Find out how many letters you can remember in the correct order.

```
N T
A N L
N F E K
B F E X A
N A Z T P L
M B T F E Q P
U N D A C X Z G
O R B V E X Z D A
R T L D C A G P V E
```

If you remembered more than seven letters in a row correctly, then you have excellent short-term memory. Your short-term memory can only hold about seven items. When you are remembering letters in a list, each letter is an 'item'. To remember more letters, chunk them into groups.

For example, the row O R B V E X Z D A has nine letters. Chunk these into groups of three: 'ORB' 'VEX' 'ZDA'.

The nine letters are easy to remember because now they are only three items. Three items doesn't overload your short-term memory.

Models of memory

The **retrieval of information**, such as word lists, is a way of testing your memory. Memory tests can tell us what memory

Summary box
- ✓ Your short-term memory can hold about seven items.
- ✓ The multistore model helps to explain how our memory works.
- ✓ The model involves sensory, short-term, and long-term memory stores.

can and cannot do. But they do not explain how the neurons in the brain work to give you memory. Explanations for how memory happens are called **models of memory**.

The multistore model: memory stores work together

Read through the list of words below once. Then cover the page and try to write down as many of the words as you can remember. They can be in any order.

> dog, window, film, menu, archer, slave, lamp, coat, bottle, paper, kettle, stage, fairy, hobby, package

How many did you remember? If this type of test is carried out on large numbers of people, a pattern is seen in the words they recall. People often remember the last few words on the list and get more of them right. They also recall the first few words on the list quite well.

When you look at a list of words:
- Nerve impulses travel from your eyes to your sensory memory.
- Some sensory information is passed on to your short-term memory. Only the information you pay attention to is passed on. You will not be able to remember words you have not noticed.
- If more information arrives than the short-term memory can hold then some is lost (forgotten). You will not remember these words either.
- Some information is passed to your long-term memory. These are words you will remember – usually the first few words on the list.
- The last information your short-term memory receives will still be there when you start to write down the list. So these are also words you will remember, usually the last few words on the list.

This use of sensory, short-term, and long-term memory stores is known as the **multistore model** of memory.

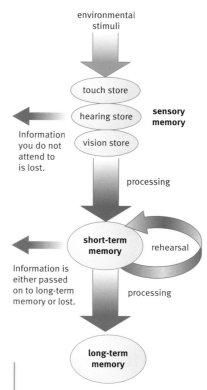

The multistore model of memory can be used to explain how some information is passed to the long-term memory store and some information is lost.

Question

5 You read the menu on a board inside a café. When you try to tell your friend sitting outside all the choices, you forget some. Why can you not remember everything on the list?

Rehearsal is one technique actors use to learn their lines.

Putting it into practice

You can apply what the psychologists have discovered to your own school work.

- *Repetition:* If you are struggling to remember a piece of information you have read, read it several times.
- *Rehearsal:* Read sections of what you have to learn that are short enough to keep in your short-term memory. Make notes from memory to help move the information to your long-term memory.
- *Active memory:* Use highlighter pens and mind maps to process information for learning.

Rehearsal helps long-term memory

Look at this row of letters:

R T I D A C G P E V

There are too many letters in this row for you to store them separately in your short-term memory. Given time you would probably repeat the letters over and over until you remembered them. **Repetition of information** is a well-known way of memorising things. An actor can memorise a sonnet (a fourteen-line poem) in around 45 minutes. Psychologists think that rehearsal moves information from your short-term memory to your long-term memory store.

The working-memory model

Rehearsed information is processed and stored rather than lost from short-term memory. You are more likely to remember information if you process it more deeply. This will happen if you understand the information or it means something to you.

For example, if you can see a pattern in the information, you process it more deeply. So:

AAT, BAT, CAT, DAT, EAT

is much easier to remember than:

DAT, AAT, EAT, CAT, BAT

You also process information more deeply if there is a strong stimulus linked to the information, for example, colour, light, smell, or sound.

An active working memory

Short-term memory is now seen as an active '**working memory**'. Here you can hold and process information that you are thinking about.

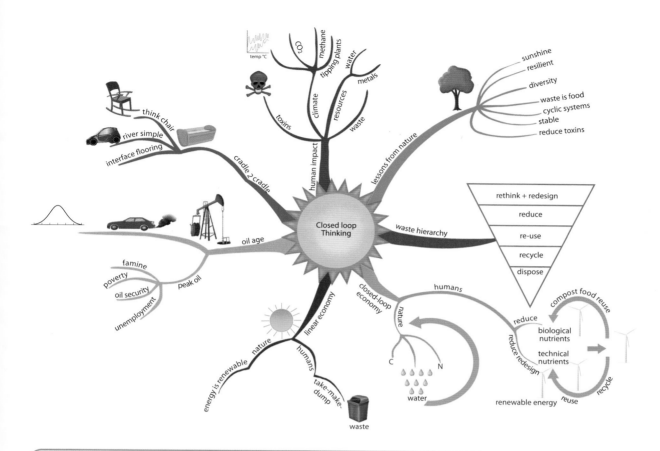

Questions

6. Make two lists of 10 different things to buy from a supermarket. Try to remember one list. Put the second list into 'families', for example tins, bakery, cleaning, and memorise it. Which list is easier to remember and why?

7. Give an example of something you can remember because a strong stimulus is linked to the information:
   a a colour   b a smell   c a sound

8. Write down an example of something you have memorised through rehearsal, for example, the directions to the cinema, or a complicated set of moves in a computer game. How has rehearsal helped you to remember this?

9. Explain why using a highlighter pen to pick out key facts makes your revision more successful.

10. Construct a visual summary like the one above to show the information you have learnt about memory and learning.

Summary box

- The working-memory model shows that rehearsal, processing, and understanding information and strong stimuli help information to be stored in our long-term memory.
- Use repetition, rehearsal, and active learning to help you remember your studies.

Science Explanations

There have been big advances in neuroscience recently, but we still do not know everything about how the human brain works. Knowing how the brain works is very important for understanding how we learn and about the mental-health diseases of old age.

You should know:

- that a stimulus is a change in the environment of an organism
- what simple reflexes are and why they are important in the role of receptors, processing centres, and effectors in body systems such as human vision
- that the body uses electrical impulses and chemical hormones for short-lived and long-lived responses, respectively
- the relationship between the central nervous system and peripheral nervous system in humans and other vertebrates
- that nerve impulses travel through relay neurons in the spinal cord and connect sensory and motor neurons
- how scientists map the regions of the cerebral cortex to particular functions
- how the evolution of a larger brain gave some early humans a better chance of survival
- the role of short-term memory and long-term memory in the storage and retrieval of information
- what helps humans to learn and recall information
- how the multistore model of memory provides a working model for short-term memory, long-term memory, repetition, storage, retrieval, and forgetting
- how simple models like the multistore model develop into more complex models such as the working-memory model.

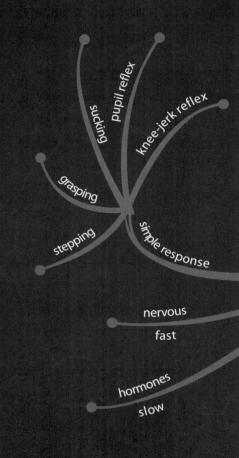

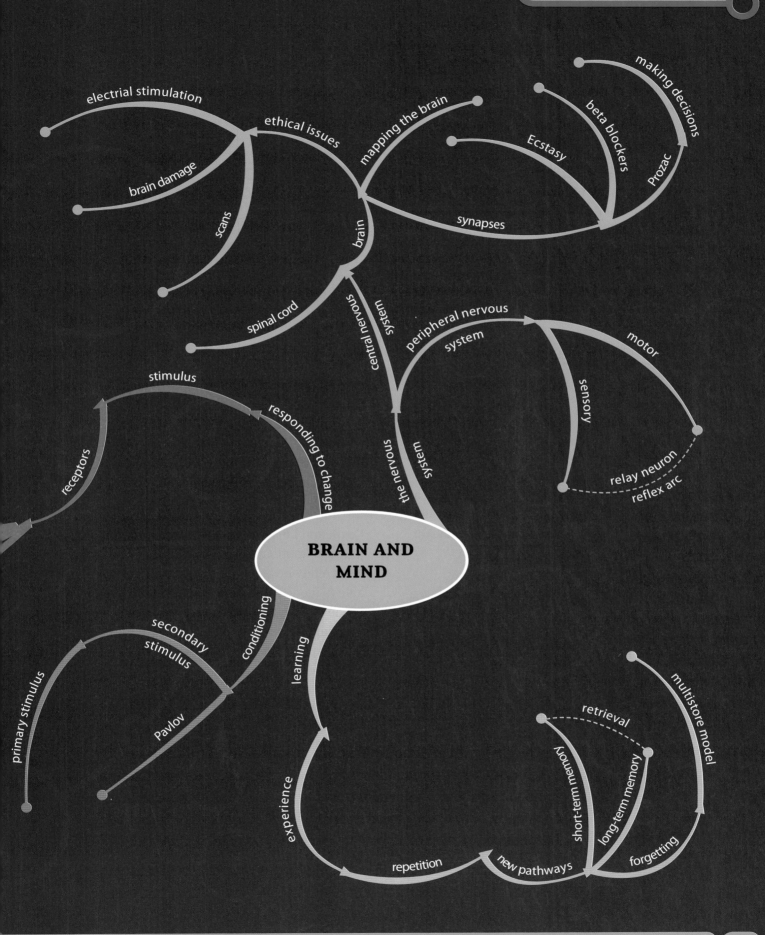

Ideas about Science

Most knowledge about the brain has come from experiments on animals and humans. This has allowed scientists to refine theories about human learning and to develop new treatments for diseases and injuries.

You need to identify ethical issues involved in scientific work and summarise the different views that may be held.
- Some people argue that using animals for medical research is acceptable if there are benefits for humans. Other people think that tampering with a vertebrate's brain cannot be justified.
- Scientists have learnt a lot about the brain from studying humans with mental-health problems. But is it fair and right to experiment on people who have an illness?
- Studies of soldiers whose brains were damaged in war have helped scientists learn how the human brain works.

It is sometimes difficult to decide what is right and what is wrong. Some people think that the right decision is one that leads to the best outcome for the greatest number of people.

Other people think that certain actions are either right or wrong whatever the consequences. You need to be able to identify arguments that are based on these two different ideas.
- With new technologies such as MRI it is possible to build up a picture of how the brain works by observing a healthy brain in action. This reduces the need for controversial experiments on animals and ill people.

This module examines some models that were developed to explain how humans remember and learn things.
- The simple multistore model is useful for only some of the observations or data on learning and memory. Fergus Craik and Robert Lockhart concluded in 1972 that the multistore memory model was too simple.

Craik and Lockhart used creative thought to produce another explanation to explain more completely how people remember. You should be able to identify where creative thinking is involved in the development of an explanation.
- The working-memory model provides a way of explaining a wider range of how people remember things.

When looking at science explanations you need to identify the better of two given scientific explanations for a phenomenon.

Review Questions

1 Here are some key terms about the nervous system. Write out the table again to match the parts correctly.

central nervous system	glands and muscles
receptors	the brain and spinal cord
synapse	peripheral nervous system
neuron	eyes, ears
effectors	the gap between two neurons
PNS	a nerve cell

2 The scientist Pavlov did research on learning using dogs. Rearrange these statements in the correct order to describe Pavlov's experiment.
  a After a while he found that the dogs would salivate even when no food was around if the bell was rung.
  b Pavlov rang a bell every time the dogs were fed.
  c Usually dogs only salivate when they see or smell food.
  d This type of learning is called conditioning.
  e The dogs had learned to link the stimulus of ringing the bell with the appearance of food.

3 Pip is a young puppy. His brain contains billions of nerve cells (neurons).
Explain what will happen to the connections between neurones (neuron pathways) in Pip's brain as he grows up.

4 Scientists can gather useful information about how the brain works by studying people with brain damage.
This research raises some ethical issues. Some people say this research is good. Others say this research is bad. Write down two different views that people might hold.

5 Name two drugs that affect the transmission of impulses across the synapses between neurons in the brain.

6 Revising for a test can be hard. Describe two different ways that you could help yourself to remember important information.

C6 Chemical synthesis

Why study chemical synthesis?

We use chemicals to preserve food, treat disease, and decorate our homes. Many of these chemicals do not occur naturally, they are synthetic (man-made). New chemical products, such as drugs to treat disease, are developed, made, and tested by chemists.

What you already know

- Atoms are rearranged during chemical reactions.
- The number of atoms of each element stays the same in a chemical reaction.
- Raw materials can be used to make synthetic materials.
- Alkalis neutralise acids to make salts.
- Some substances are made up of electrically charged particles called ions.
- Data is more reliable if it can be repeated.

Find out about

- the importance of the chemical industry
- a theory to explain acids and alkalis
- reactions that give out and take in energy
- controlling the rate of a reaction
- the steps involved in the synthesis of a new chemical
- ways to measure the efficiency of chemical synthesis.

The Science

Chemists making new chemicals need practical skills and to understand the science. They must control reactions so that they are not too slow or too fast. They must also understand if energy is given out by a reaction or needs to be put in to make it go. Acids are often used in chemical synthesis. Ionic theory can explain the reactions of these chemicals.

Ideas about Science

Technical chemists test chemicals that come from suppliers. They take repeated measurements to find the best estimate of the true value of the purity.

A · The chemical industry

Find out about

- the chemical industry
- bulk and fine chemicals
- the importance of chemical synthesis

The **chemical industry** converts raw materials, such as crude oil, natural gas, minerals, air and water, into useful products. The useful products include chemicals for use as food additives, fertilisers, pigments, dyes, paints, and pharmaceutical drugs.

The industry makes **bulk chemicals** such as ammonia, sodium hydroxide, ethene, and sulfuric acid. Bulk chemicals are made in large amounts; thousands or even millions of tonnes per year.

On a much smaller scale, the industry makes **fine chemicals** such as drugs, pesticides, and herbicides. It also makes small amounts of chemicals used by other industries to make products, for example, the liquid crystals for flat-screen televisions.

The chemical industry converts raw materials into useful products.

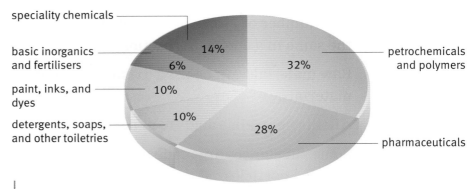

The range of products made by the chemical industry in the UK by value of sales.

The part of a chemical works that produces a chemical is called a **plant**. A chemical plant contains many different areas, with reactants and products moving from one part of the plant to another. A lot of electric power is needed for pumps to move the reactants and products around the plant. Sometimes the energy to produce this power comes from chemical reactions that give out energy.

Computer sensors monitor the conditions, such as temperature and pressure, in all the key areas of the plant.

People in the chemical industry

People with many different skills are needed in the chemical industry. Research chemists work in laboratories to find new

Questions

1 Classify the following as raw materials or products of the chemical industry: air, ammonia, water, crude oil, chlorine.
2 Give the names of two bulk chemicals.

processes and develop new products. They have to work closely with people in the marketing and sales department, who will know if the new products are wanted. If a new product is promising, it may first be tried out by making small amounts of it in a **pilot plant**.

Financial experts estimate how much the new product can be sold for. They then compare this with the cost of making the product to check that the new process will make a profit.

Once it is decided to go ahead with a project, chemical engineers design a full-scale plant. This can cost hundreds of millions of pounds. Transport workers carry the chemicals to the industry's customers.

Every chemical plant also needs managers and administrators, medical and catering staff, and training and safety officers.

Summary box

- The chemical industry converts raw materials into useful products.
- Bulk chemicals are made on a very large scale of thousands of tonnes per year.
- Fine chemicals are made on a smaller scale.
- Chemical synthesis is important because it provides chemicals for use in food, fertilisers, dyes, paints, drugs, and many other products.

Question

3 List these chemicals under two headings 'bulk chemical' and 'fine chemical':

- the acid sulfuric acid
- the drug aspirin
- the alkali sodium hydroxide
- the hydrocarbon ethene
- the herbicide glyphosate.

Plant operators monitor the plant from a control room.

Maintenance workers help to keep the plant running.

B | Acids and alkalis

Find out about

- acids and alkalis
- the pH scale
- reactions of acids

Acids

The word **acid** sounds dangerous. Nitric, sulfuric, and hydrochloric acids are very dangerous when they are concentrated. You must handle them with care because they are corrosive.

corrosive

These acids are less of a hazard when diluted with water. Dilute hydrochloric acid, for example, does not hurt the skin if you wash it away quickly, but it stings in a cut and rots clothing. It is in fact present in our stomachs where it helps to break down food and kill bacteria.

wear eye protection

All acids dissolve in water to give a solution with a pH below 7.

Ethanoic acid is a liquid. It is the acid in vinegar.

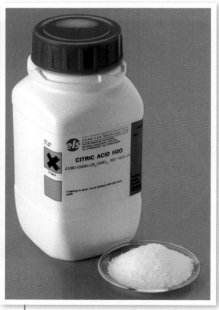
Citric acid (a solid acid) is found in citrus fruits like oranges and lemons.

Citric and tartaric acids are examples of solid organic acids. Ethanoic acid is a liquid organic acid. These acids are found in foods like oranges and vinegar.

Sulfuric, hydrochloric, and nitric acids come from mineral sources. Sulfuric acid and nitric acid are liquids at room temperature. Hydrogen chloride is a gas, and becomes hydrochloric acid when it dissolves in water.

Questions

1. Give the names of one solid acid and one liquid acid.
2. What is the pH of the solution formed when an acid dissolves in water?

Hydrogen chloride forms when concentrated sulfuric acid is added to salt (sodium chloride). Hydrogen chloride, HCl, is a gas that fumes in moist air and dissolves in water to form hydrochloric acid.

Sulfuric acid, H_2SO_4, is manufactured from sulfur, oxygen, and water. The pure, concentrated acid is an oily liquid.

Alkalis

Pharmacists sell medicines called antacids to control heartburn and indigestion. The chemicals in these medicines are the chemical opposites of acids. They neutralise excess acid produced in the stomach.

Some of these chemicals dissolve in water to give a solution with a pH above 7. Chemists call them **alkalis**. Common alkalis are sodium hydroxide (NaOH), potassium hydroxide (KOH), and calcium hydroxide ($Ca(OH)_2$).

Alkalis can do more damage to delicate tissues than dilute acids. Corrosive alkalis are used in the strongest oven and drain cleaners. They clearly have to be used with great care.

corrosive

Summary box
- Acids are chemicals that dissolve in water to form a solution with a pH below 7.
- Alkalis are chemicals that dissolve in water to form a solution with a pH above 7.

Question

3 Give the formulae of these acids and alkalis: hydrochloric acid, sulfuric acid, sodium hydroxide, calcium hydroxide.

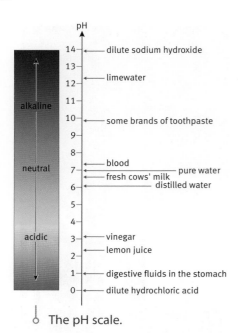

The pH scale.

A pH meter can be used to measure pH values.

Questions

4 Look at the diagram of the pH scale. State the approximate pH of the following solutions: limewater, cows' milk, vinegar, lemon juice.

5 The metal zinc reacts with hydrochloric acid to product a salt called zinc chloride and hydrogen gas. Write a word equation for this reaction.

Indicators and the pH scale

Indicators change colour to show whether a solution is acidic or alkaline. There are many different indicators.

Blue litmus turns red in acid solution and red litmus turns blue in alkalis. Special mixed indicators, such as universal indicator, show a range of colours and can be used to estimate pH values. pH values can also be measured electronically using a pH meter.

The term pH appears on many cosmetic, shampoo, and food labels. It is a measure of acidity. The **pH scale** is a number scale that shows the acidity or alkalinity of a solution. Most laboratory solutions have a pH in the range 1–14.

These hydrangea flowers contain natural indicators – they are blue if grown on acid soil and pink on alkaline soil. Note that this is the opposite of the litmus colours.

Reaction of acids

Acids with metals

Acids react with **metals** to produce **salts**. The other product is hydrogen gas.

$$\text{acid} + \text{metal} \longrightarrow \text{salt} + \text{hydrogen}$$

For example, for the reaction of hydrochloric acid with magnesium, the word equation is:

$$\text{hydrochloric acid} + \text{magnesium} \longrightarrow \text{magnesium chloride} + \text{hydrogen}$$

The balanced equation is:

$$2HCl(aq) + Mg(s) \longrightarrow MgCl_2(aq) + H_2(g)$$

Acids with metal oxides or hydroxides

An acid reacts with a **metal oxide** or **hydroxide** to form a salt and water. No gas forms.

acid + metal oxide (or hydroxide) ⟶ salt + water

For the reaction of hydrochloric acid with magnesium oxide, the word equation is:

hydrochloric acid + magnesium oxide ⟶ magnesium chloride + water

The balanced symbol equation is:
$$2HCl(aq) + MgO(s) \longrightarrow MgCl_2(aq) + H_2O(l)$$

The reaction between an acid and a metal oxide is often a vital step in making useful chemicals from ores.

Acids with carbonates

Acids react with **carbonates** to form a salt, water, and bubbles of carbon dioxide gas.

acid + metal carbonate ⟶ salt + water + carbon dioxide

Geologists can test for carbonates by dripping hydrochloric acid onto rocks. If they see any fizzing, the rocks contain a carbonate. This is likely to be calcium carbonate or magnesium carbonate.

For the reaction of hydrochloric acid with calcium carbonate, the word equation is:

hydrochloric acid + calcium carbonate ⟶ calcium chloride + water + carbon dioxide

The balanced symbol equation is:
$$2HCl(aq) + CaCO_3(s) \longrightarrow CaCl_2(aq) + H_2O(l) + CO_2(g)$$

This is a foolproof test for a carbonate. So the term 'the acid test' has come to be used to describe any way of providing definite proof.

Summary box
- The pH scale is a number scale that shows the acidity or alkalinity of a solution.
- An indicator changes colour to show if a solution is acidic or alkaline.
- Acids react with metals to form a salt and hydrogen gas.
- Acids react with metal oxides and hydroxides to form a salt and water.
- Acids react with carbonates to form a salt, water, and bubbles of carbon dioxide gas.

Questions

6 Magnesium hydroxide is an alkali used in medicines to neutralise excess stomach acid (hydrochloric acid). Write a word equation for this reaction.

7 There is a volcano in Tanzania, Africa, whose lava contains sodium carbonate. The cooled lava fizzes when added to hydrochloric acid. What gas is being produced by the reaction?

C Salts from acids

Find out about

- an ionic explanation for neutralisation reactions
- salts

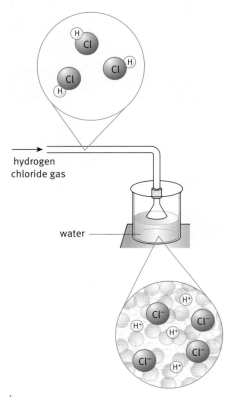

Hydrogen chloride dissolves in water to make hydrochloric acid, producing H^+ ions and Cl^- ions.

Questions

1 What ions do acids produce when they dissolve in water?

2 What ions do alkalis produce when they dissolve in water?

Explaining the properties of acids

Chemists have a theory to explain why all acids behave in a similar way when they react with indicators, metals, carbonates, metal oxides, and metal hydroxides.

Acids do not simply mix with water when they dissolve. They react, and when they react with water they produce hydrogen ions (H^+).

For example, hydrochloric acid is a solution of hydrogen chloride in water. The HCl molecules react with the water to produce **hydrogen ions** and chloride ions.

$$HCl(g) \xrightarrow{water} H^+(aq) + Cl^-(aq)$$

The theory of acids is an ionic theory. Any compound is an acid if it produces hydrogen ions when it dissolves in water.

All acids contain hydrogen in their formula. Nitric acid, HNO_3, and phosphoric acid, H_3PO_4 both contain hydrogen. But not all chemicals that contain hydrogen are acids. Ethane, C_2H_6, and ethanol, C_2H_5OH, are not acids.

Explaining the properties of alkalis

Alkalis consist of metal ions and **hydroxide** (OH^-) **ions**. When they dissolve, they add hydroxide ions to water. It is these ions that make the solution alkaline.

$$NaOH(s) \xrightarrow{water} Na^+(aq) + OH^-(aq)$$

Neutralisation

Sodium hydroxide and hydrochloric acid react to produce a salt (sodium chloride) and water. The word equation is:

sodium hydroxide + hydrochloric acid → sodium chloride + water

The balanced symbol equation showing all the ions is:

$$Na^+(aq) + OH^-(aq) + H^+(aq) + Cl^-(aq) \rightarrow Na^+(aq) + Cl^-(aq) + H_2O(l)$$

During this **neutralisation reaction** the hydrogen ions from the acid (HCl) react with hydroxide ions from the alkali (NaOH) to make water:

$$H^+(aq) + OH^-(aq) \longrightarrow H_2O(l)$$

The remaining ions in the solution make the salt. This happens in all neutralisation reactions.

Salts

Salts form when an alkali neutralises an acid. So every salt can be thought of as having two parents. Salts are related to a parent alkali and to a parent acid.

Summary box
- During a neutralisation reaction, hydrogen ions from the acid react with hydroxide ions from the alkali to make water.
- Salts form when an alkali neutralises an acid.

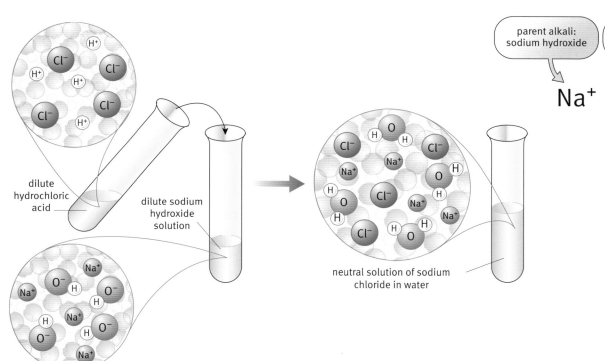

Dilute sodium hydroxide solution neutralises dilute hydrochloric acid, forming a neutral solution of sodium chloride.

Questions

3 Identify the parent acid that could be used to form these salts: lithium chloride, calcium nitrate, magnesium sulfate.

4 Identify a possible parent alkali that could be used to form these salts: sodium chloride, calcium chloride, magnesium chloride.

D — Purity of chemicals

Find out about

- purity
- titrations for testing purity

CALCIUM CARBONATE PRECIPITATED CP
QTY: 1kg   BNO: C1042/R6 - 708717

Assay	99%
Chloride (Cl)	0.005%
Sulfate (SO$_4$)	0.05%
Iron (Fe)	0.002%
Lead (Pb)	0.002%

Label on a bottle of calcium carbonate. The term 'assay' tells you how pure the chemical is. The calcium carbonate is 99% pure with small amounts of the impurities shown.

Grades of purity

The reactions of acids can be used to make valuable salts. For uses such as food or medicines, these salts have to be made pure so they are safe to swallow.

Chemicals do not always have to be pure. Limestone, for example, is used in a blast furnace to extract iron from its ores. The iron industry can use limestone straight from a quarry. Limestone has some impurities but they do not stop it from doing its job in a blast furnace.

Purifying a chemical is done in stages. Each stage takes time and money. So the higher the purity, the more expensive the chemical. Manufacturers therefore buy the quality most suitable for their purpose.

When deciding what quality of chemical to use for a particular purpose, it is important to know:
- the amount of impurities
- what the impurities are
- how they can affect the process
- whether they will end up in the product, and whether it matters if they do.

Testing purity

Chemical companies buy in many of their ingredients from chemical suppliers. Technical chemists working for the companies have to make sure that the suppliers are delivering the right quality of chemical.

For example, citric acid is often added to syrups, such as cough medicines, to control their pH. Technical chemists can check the purity of the citric acid using a procedure called a **titration**, which measures the volume of alkali that it can neutralise.

Question

1. Uses of sodium chloride salt include: **i** flavouring food, **ii** melting ice on roads, **iii** saline drips in hospitals. Put these in order of the purity of sodium chloride required, with the most pure first.

Summary box

- How pure a chemical needs to be depends on what it is used for. For use in foods or medicines, chemicals need to be very pure so that they are safe. For many other uses, chemicals do not need to be pure.
- The purity of chemicals can be tested using a procedure called a titration.

Steps involved in a titration

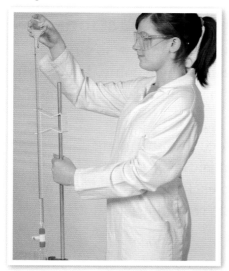

1. The technician fills a **burette** with a solution of sodium hydroxide. She knows the concentration of the sodium hydroxide solution.

2. The technician weighs out a sample of citric acid accurately.

3. The technician dissolves the acid in pure water. Then she adds a few drops of phenolphthalein indicator. The indicator is colourless in the acid solution.

4. The technician adds alkali from the burette. She swirls the flask. Near the end she adds the alkali drop by drop. At the **end point** all the citric acid is just neutralised. The indicator is now pink. The volume of alkali added is the titre.

The technician will repeat the titration several times. If there are any results that differ greatly from the rest they will be discarded. Then the mean will be calculated and the range will be found.

Questions

2. From 'steps involved in a titration' in which step is:
   a. a solution made?
   b. a pipette used?
   c. the end point reached, and how does the technician know?

3. It can be difficult to judge exactly when the colour changes and the end point is reached. Use this idea to explain why a technician repeats a titration several times.

4. 0.55 g of impure citric acid was dissolved in water and titrated with sodium hydroxide solution. At the end point 8.46 cm^3 of alkali had been added.
   a. Use this formula to find the percentage purity of the sample.

   $$\% \text{ purity} = \frac{\text{titre} \times 6.4}{\text{mass of citric acid}}$$

   b. When the titration was repeated twice more the purity was calculated to be 98.5% and 98.7%. What is the best estimate of the true value of the purity of the sample?

E Energy changes in chemical reactions

Find out about

- reactions that give out energy and reactions that take in (absorb) energy

Exothermic and endothermic reactions

Most chemical reactions need a supply of energy to get them started. Some also need energy to keep them going. That is why Bunsen burners and electric heating mantles are so common in laboratories.

Many reactions give out energy once they are going – combustion (burning), and neutralisation reactions of acids to make salts, are common examples. Reactions that give out energy to the surroundings (the surroundings get hotter) are called **exothermic**.

There are also reactions that absorb energy from the surroundings (the surroundings get cooler). These are called **endothermic**.

An electric heating mantle, used to heat reactions without a naked flame.

Both exothermic and endothermic reactions have practical uses. An exothermic reaction provides the energy for welding. A cold pack absorbs energy from an injured muscle using an endothermic reaction.

It is possible to tell whether a chemical reaction is exothermic or endothermic by measuring the temperature of the reactants before the reaction starts and the temperature of the products after it has finished. If the temperature has risen, the reaction is exothermic. If the temperature has dropped, the reaction is endothermic.

Questions

1. What happens to the temperature of a reaction mixture during an exothermic reaction?
2. What would you observe if you held a thermometer in a reaction mixture undergoing an endothermic reaction?

Energy changes in the chemical industry

Scientists working in the chemical industry need to know whether a reaction is exothermic or endothermic. This is because:

- it takes fuel, which costs money, to provide the energy needed for endothermic reactions
- the energy given out by exothermic reactions can be used elsewhere in the plant, to produce electricity, for example
- a temperature increase makes chemical reactions go faster – a reaction that gives out heat energy may get faster and faster, and can 'run away', possibly causing an explosion.

It is believed that such a 'runaway' reaction might have caused a major disaster at a chemical plant in Bhopal, India, in 1984. Thousands of people died and many people are still affected by the incident. Proper understanding of the energy changes in the reactions might have prevented the disaster.

Summary box
- Reactions that give out energy to the surroundings are called exothermic.
- Reactions that take in energy from the surroundings are called endothermic.

Some victims of the Bhopal disaster lost their sight.

Energy-level diagrams

Chemists keep track of the changes in energy in chemical reactions using **energy-level diagrams**. These show the energies of the reactants and products. Reactants are on the left and products on the right.

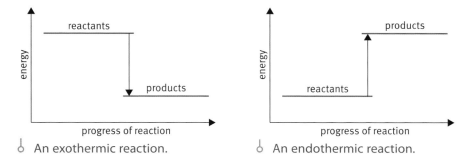

An exothermic reaction.

An endothermic reaction.

The reaction of magnesium and hydrochloric acid is an example of an exothermic reaction. Energy is given out in this reaction, so the products have less energy than the reactants.

The reaction of citric acid and sodium hydrogencarbonate is an example of an endothermic reaction. Energy is absorbed in this reaction, so the products have more energy than the reactants.

Question

3 If you have ever had a plaster cast for a broken bone, apart from the pain of the injury, you might remember a feeling of warmth as the wet plaster sets. If you put a sherbet sweet on your tongue you feel your mouth getting cold. Both of these are chemical reactions. Which is exothermic and which is endothermic?

F Rates of reaction

Find out about

- measuring rates of reaction
- factors affecting rates of reaction
- catalysts in industry
- collision theory

Rates of chemical reactions

Some chemical reactions happen very quickly. An explosion is an example of a very fast reaction.

Other reactions take time – seconds, minutes, hours, or even years. Rusting is a slow reaction and so is the rotting of food.

It is the chemist's job to work out the best way to synthesise a chemical. It is important that the chosen reactions happen at the right speed. A reaction that occurs too quickly can be dangerous. A reaction that takes too long to complete is not useful because it ties up equipment and people's time for too long. This costs money.

Measuring rates of reaction

Chemists collect data to find out how fast or slow a reaction is. They measure the **rate of a reaction** by looking at the amount of product produced or the amount of reactant used up in a fixed time.

The reaction of magnesium with hydrochloric acid produces hydrogen gas:

$$Mg(s) + 2HCl(aq) \longrightarrow MgCl_2(aq) + H_2(g)$$

The rate of this reaction can be measured quite easily by collecting and measuring the amount of hydrogen gas that is produced.

An explosion is an example of a very fast chemical reaction.

The graph on the left shows the amount of hydrogen formed over time for the reaction of magnesium with hydrochloric acid. The graph is steepest at the start, showing that hydrogen was produced quickly at the start. As the reaction continues the reaction slows down until finally it stops.

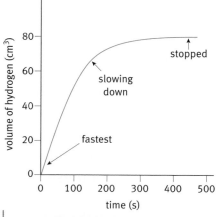

A graph showing the volume of hydrogen produced for a reaction of magnesium with hydrochloric acid.

Questions

1. Give an example of a very fast chemical reaction.
2. Give an example of a slow chemical reaction.
3. Why is it important that chemists choose reactions that happen at the right speed?

C6: CHEMICAL SYNTHESIS

Methods of measuring rates of reaction

Collecting and measuring a gas product

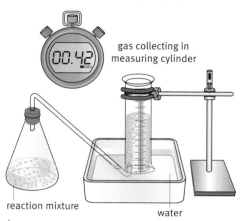

Record the volume of gas every 30 or 60 seconds. A gas syringe could be used, instead of a measuring cylinder, to collect the gas. You could use this method to measure the rate of reaction of zinc with hydrochloric acid.

Measuring the loss of mass as a gas forms

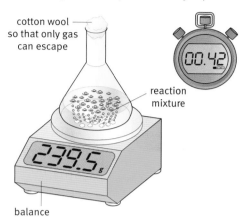

Record the mass every 30 or 60 seconds. You could use this method to measure the rate of the reaction between calcium carbonate and hydrochloric acid.

Timing how long it takes for a small amount of solid reactant to disappear

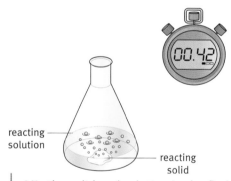

Mix the solid and solution in the flask and start the timer. Stop the timer when you can no longer see any solid. You could use this method to measure the rate of the reaction between magnesium and hydrochloric acid.

Timing how long it takes for a solution to turn cloudy

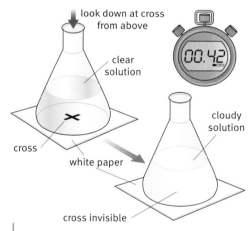

This is for reactions that produce an insoluble solid. Mix the solutions in the flask and start the timer. Stop the timer when you can no longer see the cross on the paper. You could use this method to measure the rate of the reaction between sodium thiosulfate and hydrochloric acid.

Question

4 For each of these reactions, suggest a method that could be used to measure the rate of reaction:
   a calcium carbonate reacts with dilute sulfuric acid
   b sodium thiosulfate solution reacts with dilute hydrochloric acid.

F: RATES OF REACTION

Alka Seltzer is an antacid and pain relief medicine. It comes in tablets that react in water. This reaction would be faster if the tablets were first crushed to a powder.

Factors affecting reaction rates

Powdered Alka-Seltzer medicine reacts faster in water than a tablet. Milk standing in a warm kitchen goes sour more quickly than milk kept in a refrigerator. Changing the conditions alters the rate of these processes and many others.

Factors that affect the rate of chemical reactions are:
- the *concentration* of reactants in solution – the higher the concentration, the faster the reaction
- the *surface area* of solids – powdering a solid increases the surface area and so speeds up the reaction
- the *temperature* – typically a 10 °C rise in temperature can double the rate of many reactions
- *catalysts* – these are chemicals that speed up a chemical reaction without being used up in the process.

The factors in action

The apparatus in the diagram was used in an investigation into the effect of changing the conditions on the reaction of zinc metal with sulfuric acid. The reaction produces hydrogen gas.

The graph below shows the results.

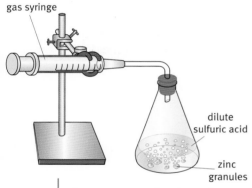

Apparatus used to investigate the reaction of zinc with dilute sulfuric acid.

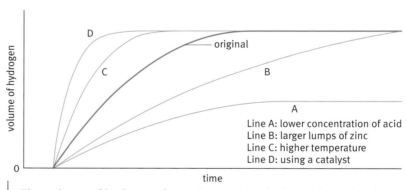

Line A: lower concentration of acid
Line B: larger lumps of zinc
Line C: higher temperature
Line D: using a catalyst

The volume of hydrogen formed over time during an investigation of the factors affecting the rate of reaction of zinc with sulfuric acid.

The red line on the graph shows the volume of hydrogen gas against time using zinc granules and 50 cm³ of dilute sulfuric acid at 20°C. The reaction gradually slows down and stops because the acid is all used up.

Each time a factor was changed, all other conditions were kept the same as in the original setup.

Question

5 When a lump of calcium carbonate is placed into a beaker of acid it reacts and carbon dioxide gas is given off. Suggest three ways to speed up the reaction.

The effect of concentration (line A)

Line A on the graph shows the results of using acid that was half as concentrated.

Halving the acid **concentration** lowers the rate of the reaction. Again, the reaction slows down over time because the concentration falls as the acid reacts with the metal. The final volume of gas is cut by half because there was only half as much acid in the 50 cm^3 of solution to start with.

The effect of surface area (line B)

Line B on the graph shows the result of using the same amount of zinc metal, but in bigger lumps. Larger lumps of metal have a smaller total **surface area** so the reaction starts more slowly. The amount of acid is unchanged and so the final volume of hydrogen is the same.

one big lump (slow reaction)

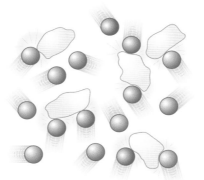
several small lumps (fast reaction)

Breaking up a solid into smaller pieces increases the total surface area. This makes it possible for the reaction to go faster.

The effect of temperature (line C)

Line C on the graph shows the result of carrying out the reaction at a higher temperature (30°C). This speeds up the reaction. The quantities of chemicals are the same so the final volume of gas collected is the same as it was originally.

The effect of adding a catalyst (line D)

Line D on the graph shows what happens when the investigation is repeated with a **catalyst** added. The reaction starts more quickly and the graph is steeper. The volume of gas at the end is the same as before.

Summary box

- There are different methods of measuring rates of reaction.
- The method that you choose depends on the reaction you are studying.
- These factors affect the rates of reactions:
  - the concentration of reactants in solution
  - the surface area of solids
  - the temperature
  - catalysts.

Questions

6 Look at the graph on the opposite page to show the reaction of zinc and sulfuric acid. Why does the reaction eventually stop?

7 When investigating the effect of temperature on a chemical reaction, why is it important to keep all other conditions the same?

Catalysts in industry

What is a catalyst?

A catalyst is a chemical that speeds up a chemical reaction. It takes part in the reaction, but is not used up.

Better catalysts

Catalysts are essential in many industrial processes. For many processes they lower the cost of making chemical products. This means the chemical products can be sold at affordable prices.

Research into new catalysts is an important area of scientific work. This is shown by the industrial manufacture of ethanoic acid from methanol and carbon monoxide.

$$\text{methanol} + \text{carbon monoxide} \longrightarrow \text{ethanoic acid}$$

$$CH_3OH(g) + CO(g) \longrightarrow CH_3COOH(g)$$

This process was first developed by the company BASF in 1960 using a cobalt compound as the catalyst. The process needed a temperature of 300 °C and a pressure 700 times atmospheric pressure.

The petrochemical company BP has now devised a new catalyst based on compounds of iridium. This process is faster and more efficient.

The new process runs at a lower temperature and a lower pressure. Iridium is cheaper, and less of the catalyst is needed. The amount of ethanoic acid produced (yield) is greater and there are fewer by-products. This makes it easier and cheaper to make pure ethanoic acid and there is less waste.

Collision theory

Chemists have a theory to explain how the various factors affect reaction rates. The basic idea is that particles can only react if they bump into each other. Imagining these particles colliding with each other can help explain the correlation between the rate of a reaction and the factors that affect it: concentration, temperature, and surface area. This is **collision theory**.

The manufacture of ethanoic acid from methanol and carbon monoxide uses a catalyst to speed up the reaction.

Questions

8 What are catalysts?

9 Why are catalysts essential to the chemical industry?

10 Suggest reasons why it is important to develop industrial processes that:
   a produce less waste
   b work at lower temperatures.

Particles are in constant motion in gases, liquids, and solutions. There are millions upon millions of collisions every second. Only a very small fraction of all the collisions are 'successful' and actually lead to a reaction.

Anything that increases the frequency of 'successful' collisions will increase the rate of reaction. Increasing the concentration of solutions of dissolved chemicals increases the frequency of collisions. The same small fraction of these collisions will be successful, but because there are more, there will also be more frequent successful collisions. Increasing surface area, by decreasing the size of solid pieces, also increases the number of 'successful' collisions.

Summary box
- Catalysts are chemicals that speed up reactions, but are not used up.
- Collision theory explains how various factors can affect reaction rates. The theory explains that particles must collide in order to react. Anything that increases the number of successful collisions will speed up the reaction.

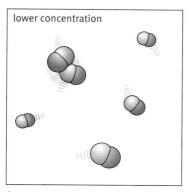
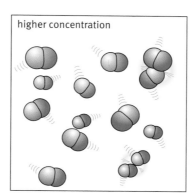

Particles have a greater chance of colliding in a more concentrated solution. Reactions get faster if the reactants are more concentrated.

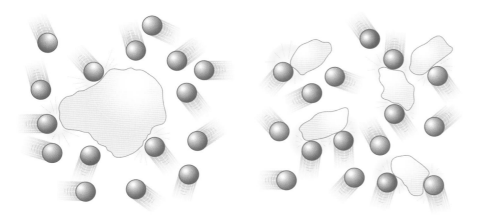

one big lump (slow reaction)   several small lumps (fast reaction)

Breaking up a solid into smaller pieces increases the total surface area. This increases the number of collisions between particles and leads to a faster reaction.

Questions

11 A scientist is studying the reaction of antacid tablets with stomach acid to see how quickly they react.
   a What is the outcome variable?
   b Give three factors that might affect the outcome variable.

12 Use collision theory to explain why increasing concentration increases the rate of a reaction.

G | Stages in chemical synthesis

Find out about

- steps in synthesis
- making a soluble salt

Chemical synthesis is a way of making new compounds. Synthesis puts things together to make something new. The stages in a chemical synthesis are:

1. Choose the reaction
2. Carry out a risk assessment
3. Carry out the reaction
4. Separate the product
5. Purify the product
6. Measure the yield and check the purity

The stages in a chemical synthesis.

An operator emptying magnesium sulfate into the tank of a sprayer on a farm.

Making a sample of magnesium sulfate

Magnesium sulfate is a salt. It is used for:
- spraying plants for healthy growth
- a raw material in soaps
- a laxative in medicine
- a supplement in feed for poultry and cattle
- a raw material in the manufacture of other compounds.

The process of making magnesium sulfate illustrates the stages in a chemical synthesis.

Stage 1: Choosing the reaction

Any of the characteristic reactions of acids can all be used to make salts:
- acid + metal ⟶ salt + hydrogen
- acid + metal oxide or hydroxide ⟶ salt + water
- acid + metal carbonate ⟶ salt + carbon dioxide + water

Magnesium metal is expensive. So it makes sense to use either magnesium oxide or carbonate as the starting point for making magnesium sulfate from sulfuric acid.

Questions

1. Which metal carbonate could be used to make the salt magnesium sulfate?
2. Which acid is needed to make the salt magnesium sulfate?

Stage 2: Carrying out a risk assessment

It is always important to minimise exposures to risk. You should take care to identify hazardous chemicals. You should also look for hazards arising from equipment or procedures. This is a **risk assessment**.

In this preparation the magnesium compounds are not hazardous. The dilute sulfuric acid is an irritant, which means that you should keep it off your skin and especially protect your eyes. You should always wear eye protection when handling chemicals.

wear eye protection

irritant

Stage 3: Carrying out the reaction

The reaction is fast enough at room temperature, especially if the magnesium oxide or carbonate is supplied as a fine powder.

This reaction can be safely carried out in a beaker. Stirring with a glass rod makes sure that the magnesium oxide or carbonate and acid mix well. Stirring also helps to prevent the mixture frothing up and out of the beaker.

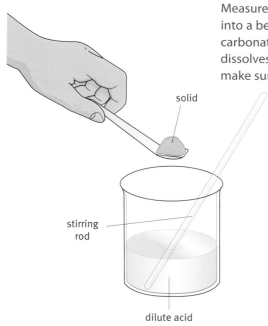

Measure the required volume of acid into a beaker. Add the metal oxide or carbonate bit by bit until no more dissolves in the acid. Warm gently to make sure all of the acid reacts.

Question

3 Write a word equation to show the reaction of magnesium oxide with sulfuric acid.

Stage 4: Separating the product from the reaction mixture

Filtering is a quick and easy way of separating the solution of the product from the leftover solid. The mixture filters more quickly if the mixture is warm.

Filter off the leftover solid, collecting the solution of the salt in an evaporating basin. The residue on the filter paper is the leftover solid.

Questions

4 Look at the procedure for making magnesium sulfate. Identify the stages that involve:
   a dissolving
   b filtering
   c evaporating
   d crystallising.

5 Look at the procedure for making magnesium sulfate and identify the stage when risks might arise from:
   a chemicals that react vigorously and spill over
   b chemicals that might spit or splash on heating
   c hot apparatus that might cause burns.

Stage 5: Purifying the product

After the mixture has been filtered, the filtrate (solution) contains the salt product dissolved in water. Evaporating much of the water speeds up crystallisation. This is carried out in an evaporating basin. The concentrated solution can then be left to cool and crystallise. The crystals are dried in an oven and then transferred to a desiccator. This is a closed container that contains a drying agent that absorbs water.

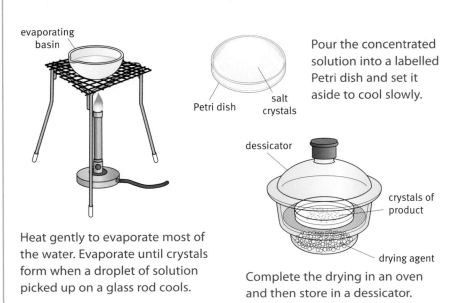

Heat gently to evaporate most of the water. Evaporate until crystals form when a droplet of solution picked up on a glass rod cools.

Pour the concentrated solution into a labelled Petri dish and set it aside to cool slowly.

Complete the drying in an oven and then store in a dessicator.

C6: CHEMICAL SYNTHESIS

Stage 6: Measuring the yield and checking the purity of the product

The final step is to transfer the dry crystals to a sample tube and weigh it to find how much product has been made. This is the actual yield of the product.

Often it is important to carry out tests to check that the product is pure. The appearance of the crystals can give a clue to the quality of the product. A microscope can help if the crystals are small. The crystals of a pure product are often well-formed and even in shape.

The weighed sample of product showing the name and formula of the chemical, the mass of product, and the date it was made.

Summary box

✓ Chemical synthesis involves a series of steps:
- choosing the reaction
- carrying out a risk assessment
- carrying out the reaction
- separating the product
- purifying the product
- measuring the yield and checking the purity.

Crystals of pure magnesium sulfate seen through a Polaroid filter (×60).

Question

6 How can the appearance of crystals of product help determine the purity of the product?

G: STAGES IN CHEMICAL SYNTHESIS

H Chemical quantities

Chemists wanting to make a certain quantity of product need to work out how much of the starting materials to use. Getting the sums right matters – especially in industry, where making more of a product for a lower price can mean better profits.

The trick is to turn the symbols in the balanced chemical equation into masses in grams or tonnes. This is possible using the relative masses of the atoms in the periodic table.

Reacting masses

Adding up the relative atomic masses for all the atoms in the formula of a compound gives the **relative formula mass** of chemicals. Given the relative formula masses, it is possible to work out the masses of reactants and products in a balanced equation. These are the **reacting masses**.

Find out about

- reacting masses
- yields from chemical reactions

RULES FOR WORKING OUT REACTING MASSES

STEP 1 Start with the balanced symbol equation.

STEP 2 Use the relative atomic masses of each atom in the formula to work out the relative formula mass of each reactant and product.

STEP 3 Write the relative formula masses under each chemical in the balanced equation.

STEP 4 Convert to reacting masses by adding the units (g, kg, or tonnes).

STEP 5 Scale the reacting masses to amounts actually used in the synthesis or experiment.

Worked example

What are the masses of reactants and products when iron (Fe) reacts with sulfur (S)?

(The relative atomic mass of Fe = 56, and S = 32)

Step 1 $Fe + S \longrightarrow FeS$

Step 2 relative formula mass of Fe = 56
relative formula mass of S = 32
relative formula mass of FeS = 88

Steps 3 & 4 Fe   +   S $\longrightarrow$ FeS
56 g   32 g   88 g

So 56 g of ion reacts with 32 g of sulfur to make 88 g of iron sulfide (FeS)

Step 5 If we started with ten times as much iron (560 g), it would react with ten times as much sulfur (320 g) to make ten times as much iron sulfide (880 g).

Questions

1 What mass of sulfur reacts with 5.6 g of iron to make 8.8 g of iron sulfide?

2 What mass of magnesium reacts with 98 g H_2SO_4 in dilute sulfuric acid?

Another way of scaling the reacting masses to the amounts actually used in a synthesis is to use a mathematical formula. For example:

Actual mass of Fe = actual mass of FeS $\times \dfrac{\text{formula mass of Fe}}{\text{formula mass of FeS}}$

So, to make 440 g of iron sulfide (FeS), the actual mass of iron (Fe) starting material required is:

Actual mass of Fe = 440 g $\times \dfrac{56}{88}$
= 280 g

Yields

The yield of any synthesis reaction is the amount of product obtained from known amounts of starting materials. The **actual yield** is the mass of the product after it has been separated from the mixture, purified, and dried.

Theoretical yield

The **theoretical yield** is the mass of product expected if the reaction goes exactly as shown in the balanced equation. This is what could be obtained in theory if there are no by-products and no losses while chemicals are transferred from one container to another. In most reactions there are some losses or by-products so the actual yield is always less than the theoretical yield.

Percentage yield

The **percentage yield** is the percentage of the theoretical yield that is actually obtained. It is always less than 100%.

> **Summary box**
> ✓ Reacting masses are the masses of reactants and products in a balanced equation.
> ✓ The yield of any synthesis reaction is the quantity of product obtained from known amounts of starting materials.

> **Worked example**
> *What is the percentage yield if 56 g of iron (Fe) produces 80 g of iron sulfide (FeS)?*
>
> From the previous example:
>
> theoretical yield = 88 g
>
> actual yield = 80 g
>
> percentage yield =
> $\dfrac{\text{actual yield}}{\text{theoretical yield}} \times 100$
>
> = $\dfrac{80 \text{ g}}{88 \text{ g}} \times 100$
>
> = 91%

> **Questions**
>
> 3 A preparation of sodium sulfate has a theoretical yield of 142 g. The actual yield of pure sodium sulfate was 128 g. Calculate the percentage yield.
>
> 4 Suggest reasons why the actual yield of a reaction is always less than the theoretical yield.

Science Explanations

Chemists use their knowledge of chemical reactions to plan and carry out the synthesis of new compounds.

You should know:

- that the chemical industry provides useful products such as food additives, fertilisers, dyestuffs, paints, pigments, and pharmaceuticals
- how chemists use indicators and pH meters to detect acids and alkalis and to measure pH
- that acidic compounds can be solids (for example, citric acid), liquids (for example, sulfuric acid), or gases (for example, hydrogen chloride)
- that common alkalis include the hydroxides of sodium, potassium, and calcium
- that there are characteristic reactions of acids with metals, metal oxides, metal hydroxides, and metal carbonates that produce salts
- how chemists use ionic theory to explain why acids have similar properties, and how alkalis neutralise acids to form salts
- why safety precautions are important when working with hazardous chemicals such as corrosive acids and alkalis
- how to follow the rate of a change by measuring the disappearance of a reactant or the formation of a product and then to analyse the results graphically
- that the concentrations of reactants, the particle size of solid reactants, the temperature, and the presence of catalysts are factors that affect the rates of reaction
- how collision theory can explain why changing the concentration of reactants, or the particle size of solids, affects the rate of a reaction
- that reactions are exothermic if they give out energy and endothermic if they take in energy
- that a chemical synthesis involves a number of stages and a range of practical techniques, which are important in achieving a good yield of a pure product in a safe way
- that a titration is a procedure that can be used to check the purity of chemicals used in synthesis.

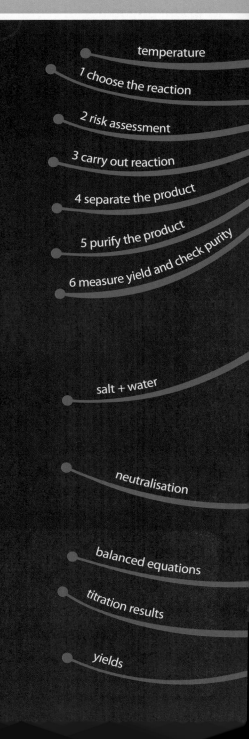

temperature
1 choose the reaction
2 risk assessment
3 carry out reaction
4 separate the product
5 purify the product
6 measure yield and check purity
salt + water
neutralisation
balanced equations
titration results
yields

Ideas about Science

Scientists can never be sure that a measurement tells them the true value of the amount being measured. Data is more reliable if it can be repeated. If you take several measurements of the same quantity, for example, in a titration, the results are likely to vary. This may be because:
- the quantity you are measuring is varying, for example, the purity of separate solid samples of a product may be different
- different people have different ideas about when an indicator has changed colour and the end point of a titration has been reached, or of the position of a meniscus in a burette
- different types of measuring equipment have different levels of accuracy, for example, some balances measure to more decimal places than others.

Usually the best estimate of the true value of an amount is the mean (or average) of several repeat measurements. You should:
- be able to calculate the mean from a set of repeat measurements

- know that a measurement may be an outlier if it lies well outside the range of the other values in a set of repeat measurements
- treat an outlier as data unless there is a reason to doubt its accuracy.

To investigate the relationship between a factor and an outcome, it is important to control all the other factors that might affect the outcome, such as temperature and the concentrations of other reactants.

When investigating the rates of chemical reactions you should be able to:
- identify the outcome and factors that may affect it
- suggest how an outcome might alter when a factor is changed.

In a plan for an investigation of the effect of a factor on an outcome, you should:
- recognize that controlling other factors is a positive design feature, and that not controlling other factors is a design flaw.

If an outcome occurs when a specific factor is present but does not occur when it is absent, or if an outcome variable increases (or decreases) steadily as an input variable increases, we say that there is a correlation between the two.

In the context of studying reaction rates you should:
- be able to identify where a correlation exists when data is presented as text, as a graph, or in a table
- understand that a correlation does not always mean that the factor causes the outcome
- identify that where there is a machanism to explain a correlation, scientists are likely to accept that the factor causes the outcome.

Review Questions

C6: CHEMICAL SYNTHESIS

1 Some students are researching how the energy changes of reactions may be used.
  a For each student name the type of reaction needed.
   i A student is developing a new sports pack. She needs to find two chemicals that will cool an injury when mixed inside the pack.
   ii Another student is developing a hand-warming pack to place inside gloves. He needs to find two chemicals that will warm his hands when mixed inside the pack.
  b The first student thinks that the chemicals will 'make cold'. Explain why this is not the case.

2 A chemist tested the purity of a sample of tartaric acid by titrating with dilute sodium hydroxide. The table shows his results.

	Run 1	Run 2	Run 3	Run 4
Initial burette reading (cm^3)	0.1	0.1	25.0	21.9
Final burette reading (cm^3)	25.2	25.0	50.0	46.9
Volume of alkali added (cm^3)	25.1	24.9	25.0	

  a Explain why the chemist repeated the titration four times.
  b Calculate the volume of alkali added in Run 4.
  c The chemist calculates the mean of his titration results. Why does he do this?
  d Suggest the range within which the true value probably lies.

3 A teacher adds 3 g of zinc granules to dilute hydrochloric acid in a flask. She uses a gas syringe to measure the volume of hydrogen gas given off.

Time (min)	Volume of gas in syringe (cm^3)
1	32
2	56
3	74
4	87
5	95
6	95

  a Draw a line graph of the results.
  b How long does it take for the reaction to finish?
  c When the reaction finishes there is still some zinc left in the flask. Suggest why.
  d The teacher decides to repeat the experiment with 3 g of zinc powder.
   i Predict how this will affect the rate of reaction.
   ii Draw a dotted line on the graph to show the expected results of this second experiment.

4 Zinc sulfate is a soluble salt used in some dietary supplements. Zinc sulfate can be made by gradually adding zinc oxide to an acid.
  a Which acid could be added to zinc oxide to make zinc sulfate?
  b What would be observed when the zinc oxide is first added to the acid?
  c What observation shows that all the acid has reacted with the zinc oxide?
  d Describe how to obtain a pure sample of zinc sulfate from this solution.

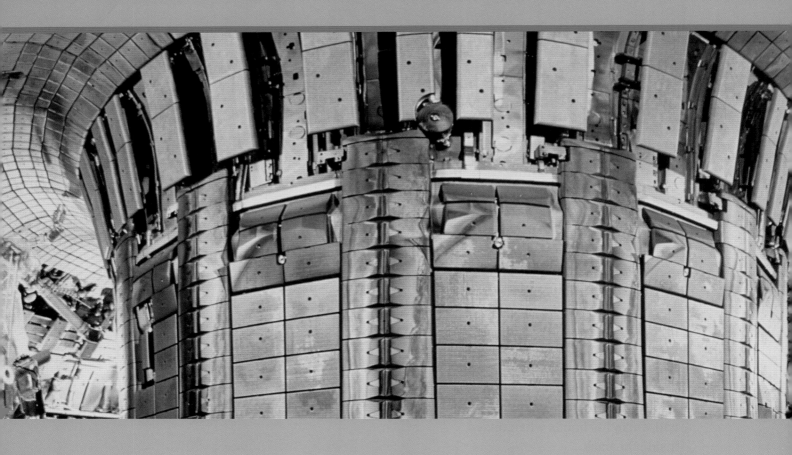

P6 Radioactive materials

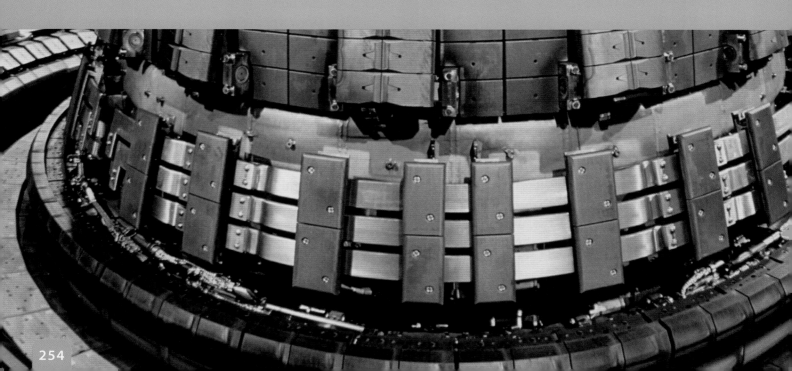

Why study radioactive materials?

People make jokes about radioactivity. If you have hospital treatment with radiation, they may say you will 'glow in the dark'. People may worry about radioactivity when they don't need to. Most of us take electricity for granted. But today's power stations are old. Should nuclear power stations be built as replacements? Should we research nuclear fusion as a long-term energy solution?

What you already know

- Some materials are radioactive, and naturally emit gamma rays.
- Gamma rays are ionising radiation.
- Ionising radiation can damage living cells.
- Nuclear power stations produce radioactive waste.
- Contamination by a radioactive material is more dangerous than a short period of irradiation.

Find out about

- radioactive materials and emissions
- radioactive materials being used to treat cancer
- ways of reducing risks from radioactive materials
- nuclear power stations and fusion research.

The Science

The discovery of radioactivity changed ideas about matter and atoms. The nuclear model of the atom helped scientists explain many observations – including radioactivity and the colour of stars.

Ideas about Science

Physicists' understanding of radioactivity has enabled them to develop many applications – from nuclear power stations to cancer detection and treatment. Knowledge of radioactivity is essential to develop safe ways of working.

A | Radioactive materials

Find out about

- radioactive decay
- what makes an atom radioactive
- types of radiation

What do these elements have in common?

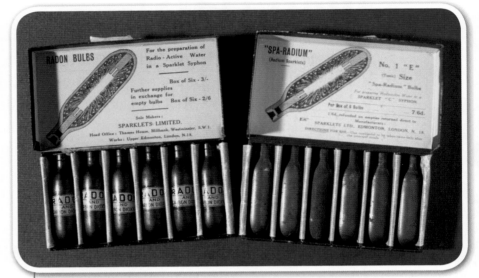

Radon is a radioactive gas. Radium is a radioactive metal. In the early 1900s, these bulbs were used to make drinking water radioactive.

Uranium ore – uranium is used as a nuclear fuel in power stations.

This scientist is measuring the radioactivity of the fruit from plants grown with radioactive water.

They are all **radioactive**. If you test them with a Geiger counter you will hear it click.

When radioactivity was first discovered, people did not know that the radiation was **ionising** and could damage or kill living cells. They thought that it was natural and healthy. Manufacturers made all kinds of products using radioactive materials. When scientists realised the danger the products were banned. Safety rules for using radioactive materials were introduced.

Radioactive elements can be naturally occurring, or they can be man-made. The man-made elements may be produced because they are useful. For example, the radioactive hydrogen in the water molecules is used to water the plants in the photograph. They may be a waste product, like waste from nuclear power stations.

P6: RADIOACTIVE MATERIALS

Changes inside the atom

Many elements have more than one type of atom. For example, some carbon atoms are radioactive. In most ways they are identical to other carbon atoms. All can:
- be part of coal, diamond, or graphite
- be a part of molecules
- take point in chemical reactions, for example, burn to form carbon dioxide.

Radioactive decay

The main difference is that most carbon atoms do not change. They are stable.

Radioactive carbon atoms randomly give out energetic radiation. Each atom does it only once. And what is left afterwards is not carbon, but a different element . The process is called **radioactive decay**. It is not a chemical change; it is a change *inside* the atom.

What makes an atom radioactive?

Atoms have a tiny core called the **nucleus**. In some atoms, the nucleus, is **unstable**. The atom decays to become more stable. It emits energetic radiation and the nucleus changes. This is why the word 'nuclear' appears in *nuclear reactor*, *nuclear medicine*, and *nuclear weapon*.

Three types of radiation are emitted, called alpha, beta, and gamma.

A cut diamond sitting on a lump of coal. Each of these is made of carbon atoms. Some of the atoms will be radioactive.

Radiation	What it is
alpha particles (α)	small, high speed particle with + charge
beta particles (β)	smaller, higher speed particle with − charge
gamma radiation (γ)	high energy electromagnetic radiation

Questions

1. How can you test to see if something is radioactive?
2. Why is ionising radiation dangerous?
3. What are the three different types of radiation from radioactive materials?

Summary box

✓ Some atoms are radioactive. They decay by emitting alpha particles or gamma radiation.

A: RADIOACTIVE MATERIALS

B | Atoms and nuclei

Find out about

- models of the atom
- how alpha particle scattering reveals the existence of the atomic nucleus

How do scientists know about the structure of atoms?

The 'solar system' model of the atom dates back to 1910, and an experiment thought up by Ernest Rutherford. Scientists were beginning to understand radioactivity, and were experimenting with radiation. Rutherford realised that alpha particles were smaller than atoms, and so they might be useful tools for investigating the structure of atoms. So he designed a suitable experiment, and it was carried out by his assistants, Hans Geiger and Ernest Marsden.

Here is how to do it:
- Start with a metal foil. Use gold, because it can be rolled out very thin, to a thickness of just a few atoms.
- Direct a source of alpha radiation at the foil. Do this in a vacuum, so that the alpha particles are not absorbed by air.
- Watch for flashes of light as the alpha particles strike the detecting material on the screen at the end of the microscope.
- Work all night, counting the flashes at different angles, to see how much the alpha radiation is deflected.

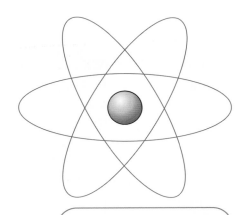

The 'solar system' model of the atom has a nucleus at the centre and electrons whizzing around like miniature planets.

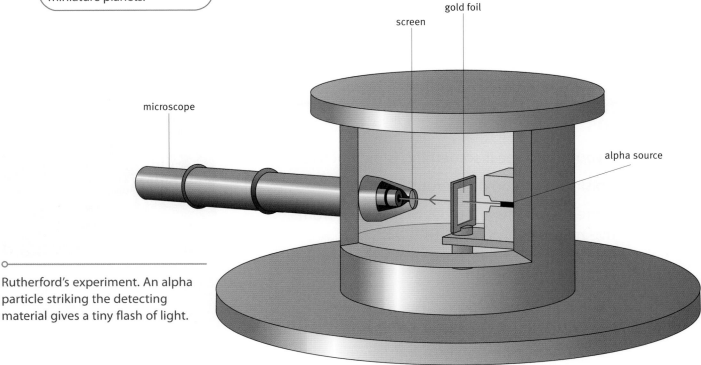

Rutherford's experiment. An alpha particle striking the detecting material gives a tiny flash of light.

Results and interpretation

This is what Geiger and Marsden observed:
- Most of the alpha particles passed straight through the gold foil, deflected by no more than a few degrees.
- A few of the alpha particles were actually reflected back towards the direction from which they had come.

And here is what Rutherford said:
'It was as if, on firing a bullet at a sheet of tissue paper, the bullet were to bounce back at you!'

In fact, very few alpha particles were back-scattered, but it still needed an explanation.

Rutherford realised that there must be something with positive charge repelling the alpha particles because they also have positive charge. And it must also have a lot of mass, or the alpha particles would just push it out of the way.

This 'something' is the nucleus of a gold atom. It contains all of the positive charge within the atom, and most of the mass. Rutherford's nuclear model is a good example of a scientist using creative thinking to develop an explanation of the data.

His analysis of his data showed that the nucleus was very tiny, because most alpha particles flew straight past without being affected by it. The diameter of the nucleus of an atom is roughly a hundred-thousandth of the diameter of the atom.

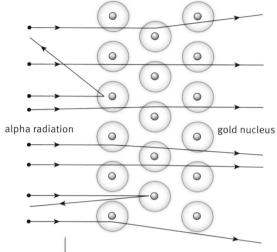

Only alpha particles passing close to a nucleus are deflected by a large amount.

Questions

1. What charge do the following have:
   a the atomic nucleus?
   b alpha radiation?
   c electrons?

2. Put these in order, from least mass to greatest: gold atom, alpha particle, gold nucleus, electron.

3. Describe and explain what happened to alpha particles that were directed:
   a straight towards a gold nucleus
   b slightly to one side of a gold nucleus
   c midway between two nuclei.

Summary box
- The alpha particle experiment should that atoms have a small massive nucleus with positive charge.
- Rutherford used creative thinking to develop the nuclear model explanation.

C Inside the atom

Find out about
- **protons and neutrons**
- α and β particles

Atoms are small – about a ten millionth of a millimetre across. Their outer layer is made of electrons. Most of their mass is concentrated in a tiny core, called a nucleus.

The nucleus

The tiny nucleus contains two types of particle: **protons** and **neutrons**. Carbon-11 and carbon-12 are different forms of carbon. Carbon-12 has 12 particles in the nucleus: six protons and six neutrons. Carbon-11 has 11 particles in the nucleus: six protons and five neutrons. These different forms are called isotopes of carbon.

Carbon-11 will give out its radiation whether it is in diamond, coal, or graphite. You can burn it or vaporise it and it will still be radioactive.

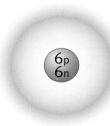

carbon-12

carbon-11

Carbon-11 has 11 particles in its nucleus: 6 protons and 5 neutrons. The nucleus of carbon-12 has 6 protons and 6 neutrons.

Compared to the whole atom, the tiny nucleus is like a pinhead in a stadium.

Making gold

When radioactive platinum decays it turns into a new element – gold. A good way to make money?
No. The price of gold is only half the price of platinum.

beta radiation

Radioactive changes

Some nuclei that are unstable can become more stable by emitting an alpha particle.

Other nuclei can become more stable by emitting a beta particle. These particles come from the nucleus of the atom.

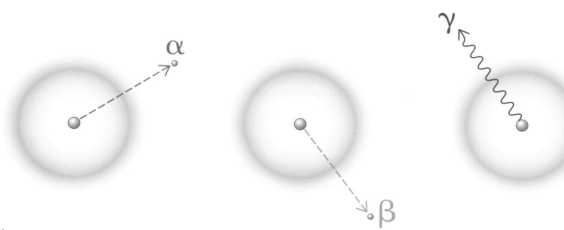

It is the nucleus of an atom that makes it radioactive and emits the radiation.

The emission of either an alpha or a beta particle from an unstable nucleus produces a nucleus of a different element, called a 'daughter product' or 'decay product'. The daughter product may itself be unstable. There may be a series of changes, but eventually a stable end-element is formed.

Questions

1 Look at these nuclei:
   A carbon-11
   B boron-11
   C carbon-12
   D nitrogen-12

   a Which two are the same element?
   b Which ones have the same number of particles in the nucleus?
   c Do any of them have identical nuclei?

2 Put in order of size with the biggest first proton, atom, nucleus, molecule, pinhead.

3 What part of the atom does the radiation come from?

Summary box
- Alpha or beta particles are emitted by an unstable nucleus.
- This produces a nucleus of a different element.

D | Using radioactive isotopes

Find out about
- alpha, beta and gamma radiation
- sterilisation using ionising radiation

Radioactive isotopes have many uses, but they are quite rare in Nature – because most of them have decayed – so radioactive isotopes are made in nuclear reactors for use in laboratories and hospitals around the country.

Alpha, beta or gamma?

To decide which radiation to use, scientists consider these properties.

Alpha radiation
Alpha particles are much heavier than beta particles, and they quickly collide with air molecules and slow down.

This means that they are the least penetrating, but also the most strongly ionising radiation. They are stopped most easily.

Beta radiation
Beta particles move very fast. They are much smaller than alpha particles so less likely to collide with other particles.

This means they travel further in air and other materials and are less ionising.

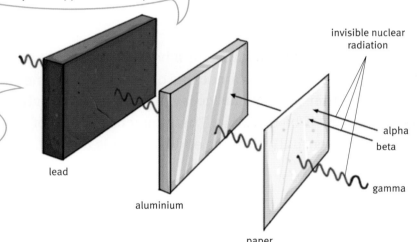

Uses of ionising radiations are linked to their properties.

Gamma radiation
Sometimes, the protons and neutrons in the nucleus just rearrange themselves to become more stable. When this happens the nucleus emits a photon of electromagnetic radiation called a gamma ray.

This does not cause a change of element. The photons have more energy than most X-ray photons and rarely collide with particles, so the radiation is very penetrating. It has only a very weak ionising effect.

Question

1 Which type of radiation:
   a is the most penetrating?
   b is the most ionising?
   c has the longest range in air?

Properties of ionising radiation

radiation	range in air	stopped by	ionisation	charge
alpha	a few cm	paper / dead skin cells	strong	+
beta	10–15 cm	thin aluminium	weak	–
gamma	many metres	thick lead or several metres of concrete	very weak	no charge

Sterilisation

Ionising radiations can kill bacteria. Gamma radiation is used for sterilising surgical instruments and some hygiene products such as tampons.

- The products are first sealed from the air and then exposed to the radiation.
- This passes through the sealed packet and kills the bacteria inside.

Food can be treated in the same way. Irradiating food kills bacteria so the food lasts longer. Since 2010, **irradiation** is permitted in the UK for herbs and spices. The label must show that they have been treated with ionising radiation. This can be better than heating or drying, because it does not affect the taste.

Questions

2 a Why seal the packets of surgical instruments before sterilising them?
   b Does the gamma radiation make them radioactive? Explain your answer.
3 Smoke detectors used in homes contain a source that emits alpha particles.
   a Explain why these are not dangerous in normal use.
   b What might make them dangerous?

Food irradiation is it safe?

The logo shows that the herbs and spices have been **irradiated** with gamma radiation from Cobalt–60. Gamma rays pass through the glass and kill any bacteria in the jar. Cobalt–60 does not pass in to the jar– there is no **contamination**.

Uses of ionising radiation are linked to their properties.

Gamma rays kill the bacteria on and inside these test tubes.

Summary box

- Alpha, beta, and gamma radiations have the different properties described above.
- Gamma radiation is used to sterilise food and surgical products.

E | Radiation all around

Find out about

- background radiation
- a radioactive gas called radon
- radiation dose and risk

Radiation sources

If you switch on a Geiger counter, you will hear it click. It is picking up **background radiation**, which is all around you. Most background radiation comes from natural sources.

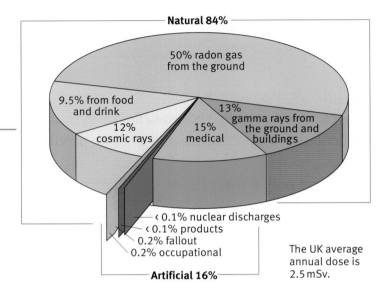

How different sources contribute to the average **radiation dose** in the UK. Source HSE

The UK average annual dose is 2.5 mSv.

Radiation dose

Radiation dose measures the possible harm the radiation could do to the body. It is measured in millisieverts (mSv).

- The UK average dose is 2.5 mSv a year.
- With a dose of 1000 mSv (400 times larger) three out of a hundred people, on average, develop a cancer.

Ionising radiation from outer space is called cosmic radiation.

- Flying to Australia gives you a dose of 0.1 mSv, from cosmic rays. That's not much if you go on holiday, but it soon adds up for flight crews.

What affects radiation dose?

The dose measures the possible harm done by the radiation. It depends on:

- the amount of radiation reaching the body
- the type of radiation. Alpha is the most ionising of the three radiations. So it can cause the most damage to a cell. The same amount of alpha radiation gives a bigger dose than beta or gamma radiation.

Questions

1. **a** In what units is radiation dose measured?
   b What is the average radiation dose per year in the UK?
   c What percentage of the average radiation dose in the UK comes from food and drink?

2. **a** How big a dose of radiation do you get by catching a flight to Australia?
   b Where do cosmic rays come from?

The damage to the body depends on the type of tissue affected. Lung tissue is easily damaged. Radon gas is dangerous because it emits alpha particles. If it is breathed into the lungs then the alpha radiation will be absorbed in the lung tissue.

Is there a safe dose?

There is no such thing as a safe dose. Just one radon atom might cause a cancer. This is like a person being knocked down by a bus the first time they cross a road. The chance of it happening is low, but it still exists. The lower the dose, the lower the risk. But the risk is never zero.

Irradiation

Irradiation is when you are exposed to a radiation source outside your body. Alpha irradiation presents a very low risk because alpha particles:
- only travel a few centimetres in air
- are easily absorbed.

Your clothes will stop alpha particles. So will the outer layer of dead cells on your skin.

Irradiation by beta particles is more risky as they penetrate a few centimetres into the body.

Most gamma rays pass straight through the body. They have high energy, so if they are absorbed they are dangerous.

Contamination

Contamination is when a radiation source enters your body, or gets on your skin or clothes. You become contaminated. If you swallow or breathe in any radioactive material, your organs will be exposed to continuous radiation. Sources that emit alpha particles are the most dangerous because alpha particles are the most ionising. Contamination by gamma sources is the least dangerous as most gamma rays will pass straight out of the body.

In the 1970s, Alice Stewart's research suggested that radiation is more harmful to children and to elderly people. She was attacked for her ideas.

Summary box
- Radiation dose is affected by:
  - amount of radiation
  - type of radiation.
- Radiation dose is measured in millisieverts (mSv).

Radiation protection

Health physicists study radiation hazards and give advice to protect against them. They also keep a close eye on people who work with radioactive materials, for example, in hospitals, and industry. These people are called 'radiation workers'.

Employers must ensure that radiation workers receive a radiation dose 'as low as reasonably achievable'.

For example, if better equipment would reduce the risk, and the cost is reasonable, they must buy it.

There is guidance to protect hospital patients who receive radiation treatment too. If one hospital uses smaller doses but its results are just as good, then all hospitals are asked to copy them.

Working with radiation

Manisha is a nuclear medicine technician in a hospital. She prepares radioactive isotope doses and may be exposed to radiation. To keep her dose low she:
- uses protective clothing and screens
- wears gloves and an apron
- wears a personal radiation monitor whenever she is working.

Staff handle radioactive sources with gloves and forceps.

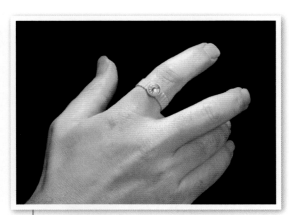

People working with radiation wear a personal radiation monitor to keep track of their dose.

Questions

3 On what two factors does radiation dose depend?

4 Explain the difference between irradiation and contamination.

5 Explain how each of the precautions Manisha takes helps her to keep her radiation dose as low as possible.

Living with radon

Radon gas

Over 400 years ago, a doctor wrote about the high death rate amongst German silver miners. He thought they were being killed by dust, causing disease.

We now know that radon gas is harmful because it is **radioactive**. It produces **ionising radiation** that can damage cells. The silver miners were dying of lung cancer.

Radon and lung cancer

Radon seeps into houses in some areas of the UK, as described in the leaflet on the following page. Scientists have done lots of studies to see if a low dose over a long time increases the risk of lung cancer.

Scientists measure radon levels in the homes of people with lung cancer and compare them with levels in homes of people who have not got lung cancer. One study:
- chose women who had lived in the same homes for 20 years
- compared 413 women with lung cancer with 614 without lung cancer
- showed a link between radon exposure and lung cancer.

Some studies have not shown a link. This may be because:
- they had a smaller sample size
- it is difficult to measure radon exposure over time, especially if people move.

Find out about
- radon gas
- radiation dose and risk

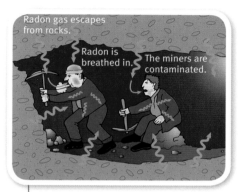

Silver mines were contaminated with radon gas. The miners breathed it in and suffered.

The pipe runs beneath the floor of the house and a small fan sucks the radon from the building.

Question

1 a What was the correlation that the doctor observed?

b What was i) the factor and ii) the outcome in what happened to the silver miners?

A hazard at home

Radon gas builds up in enclosed spaces. In some parts of the UK, it seeps into houses.

LIVING WITH RADON

GOVERNMENT INFORMATION LEAFLET

There is radon all around you. It is radioactive and can be hazardous – especially in high doses.

Radon gives out a type of ionising radiation called **alpha radiation**. Like all ionising radiations, alpha radiation can damage cells and might start a cancerous growth.

Radon is a gas that can build up in enclosed spaces. Some homes are more likely to be contaminated with radon.

What about my home?

You and your family are at risk if you inhale radon-contaminated air. The map shows the areas where there is most contamination.

If you live in one of these areas, get your house tested for radon gas.

What if the test shows radon?

Radon comes from the rocks underneath some buildings. It seeps into unprotected houses through the floorboards. If your house is contaminated, get it protected. An approved builder will put in:
- a concrete seal to keep the radon under your floorboards and
- a pump to remove it safely.

The risk is real: put in a seal.

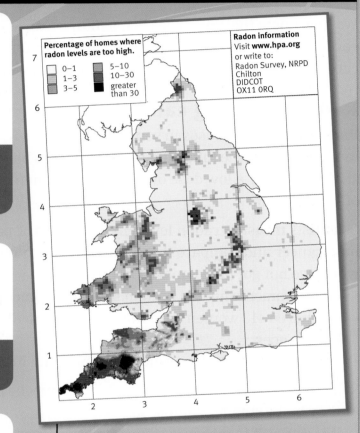

Radon-affected areas in England and Wales. Based on measurements made in over 400 000 homes.

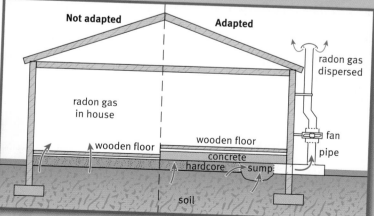

Radon gas can build up inside your home. Sealing the floor and pumping out the gas is an effective cure.

Radon and risk

The risk to miners was high because radon can build up in enclosed spaces, such as mines. In the atmosphere, the radon spreads out. In mines the rocks keep producing the gas and it cannot escape. So the radon concentration can be 30 000 times higher than in the atmosphere.

There is a lower concentration in a house. And it is much lower if the windows are open or there are draughts.

On average, radon makes up half the UK annual radiation dose. About 1100 people die each year from its effects. That is about 1 in every 50 000 people. Radon is only one hazard. There are risks with driving to school, sunbathing, swimming, and even eating.

Many risky activities have a benefit. You need to decide whether to take the risk.

The table shows how the risk of cancer from radon compares with some other common risks.

Cause of death	Average number of deaths per year
cancer caused by radon	1100
cancer among workers caused by asbestos	4000
skin cancer caused by ultraviolet radiation	1400
road deaths	2500
cancer caused by smoking	35 000
CJD	98
house fire	360
all causes	510 000

Estimated deaths per year in the UK population of 60 million (2008).

Questions

2 There is a risk from radon gas building up in houses. Which of these are good ways to reduce the risk?
- stop breathing
- wear a special gas mask
- move house
- adapt the house.

3 Choose three causes of death from the table on the left. Write down a way of reducing the risk from each one.

4 Write a letter to a friend living in a high-radon area to persuade them to get their house checked for radon.

Summary box
- Ionising radiation can damage living cells. This may cause cancer.
- Radioactive materials may irradiate or contaminate people.
- The risk depends on the radiation dose.

G Half-life

Find out about
- the half-life of radioactive materials

Radioactive decay is random. You can never tell which nucleus will decay next. Scientists can't predict whether a particular nucleus will decay today or in a thousand years time. But in a sample of radioactive material there are billions of atoms, so they can see a pattern in the decay.

The pattern of radioactive decay

The amount of radiation from a radioactive material is called its **activity**. This decreases with time.
- At first there are a lot of radioactive atoms.
- Each atom gives out radiation as it decays to become more stable.
- The activity of the material falls because fewer and fewer radioactive atoms remain.

Half-life

Technetium-99m is a radioactive element used as a medical tracer. The diagram shows what happens when it is injected into a patient at 9:00 am.

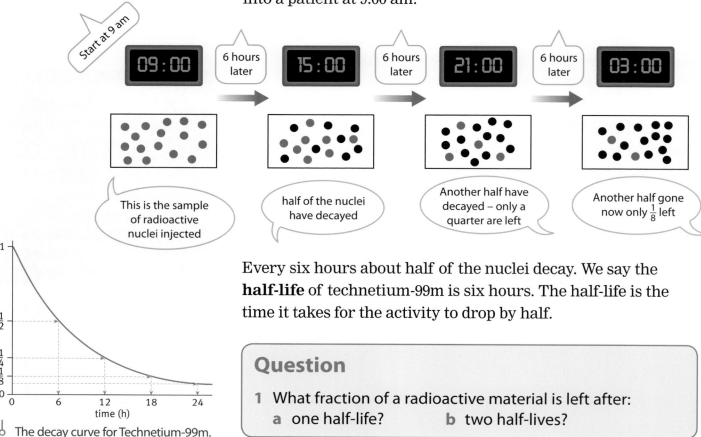

Start at 9 am

09:00 → 6 hours later → 15:00 → 6 hours later → 21:00 → 6 hours later → 03:00

This is the sample of radioactive nuclei injected

half of the nuclei have decayed

Another half have decayed – only a quarter are left

Another half gone now only $\frac{1}{8}$ left

The decay curve for Technetium-99m.

Every six hours about half of the nuclei decay. We say the **half-life** of technetium-99m is six hours. The half-life is the time it takes for the activity to drop by half.

Question

1 What fraction of a radioactive material is left after:
   a one half-life?   b two half-lives?

Time	Hours since injection	Number of half-lives	Fraction of original sample remaining
9.00 am	0	0	1
3.00 pm	6	1	$\frac{1}{2}$
9.00 pm	12	2	$\frac{1}{4}$
3.00 am	18	3	$\frac{1}{8}$
9.00 am	24	4	$\frac{1}{16}$

The radioactive decay of technetium-99m.

The six-hour half-life makes it a useful medical tracer. It lasts long enough for doctors to get some scans of the decay, but it has almost all gone in a few days.

Different half-lives

All radioactive materials show the same pattern but they can have different half-lives. The graph on the right shows the pattern of radioactive decay for radon.

There is no way of slowing down or speeding up the rate at which radioactive materials decay. No chemical reaction of physical change makes any difference. Some decay slowly over thousands of millions of years. Others decay in milliseconds – less than the blink of an eye.

The shorter the half-life, the greater the activity for the same amount of material. Of the four radioactive isotopes listed in the table on the right, neon-17 is the most active.

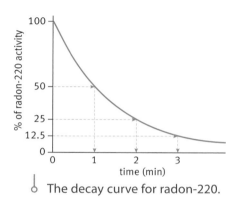

The decay curve for radon-220.

Isotope	Half-life
Iridium-192	74 days
Strontium-81	22 minutes
Uranium-235	710 million years
Neon-17	0.1 seconds

Half-lives can be short or long.

Question

2 Iodine-123 is used to investigate problems with the thyroid gland. It is a gamma emitter.
   a Explain why it is useful that iodine-123 gives out gamma radiation.
   b Iodine-123 has a half-life of 13 hours. Why would it be a problem if the half-life was:
      i a lot shorter?
      ii a lot longer?

Summary box
- **Half-life is the time it takes for the activity of a radioactive material to drop by half.**
- **Radioactive materials have a wide range of half-lives.**

H — Medical imaging and treatment

Find out about

- different uses of radiation
- types of radiation
- benefits and risks of using radioactive materials
- limiting radiation dose

Radioactive materials can cause cancer. But they can also be used to diagnose and cure many health problems.

Medical imaging

Jo has been feeling unusually tired for some time. Her doctors decide to investigate whether an infection may have damaged her kidneys when she was younger.

They plan to give her an injection of DMSA. This is a chemical that is taken up by normal kidney cells.

The DMSA has been labelled as radioactive. This means its molecules contain an atom of technetium-99m (Tc-99m), which is radioactive.

The Tc-99m gives out its gamma radiation from within the kidneys. Gamma radiation is very penetrating, so nearly all of it escapes from Jo's body and is picked up by a gamma camera.

Jo's scan shows that she has only a small area of damage. The doctors will take no further action.

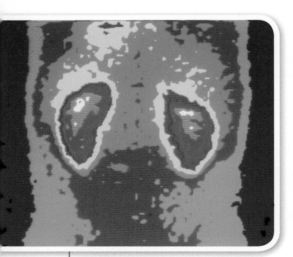

This gamma scan shows correctly functioning kidneys – the top two white areas.

Glowing in the dark

Jo was temporarily contaminated by radioactive technetium. For the next few hours, until her body got rid of the technetium, she was told to:
- flush the toilet a few times after using it
- wash her hands thoroughly
- avoid close physical contact with friends and family.

Is it worth it?

There was a small chance that some gamma radiation would damage Jo's healthy cells. Before the treatment, her mum had to sign a consent form, and the doctors checked that Jo was not pregnant.

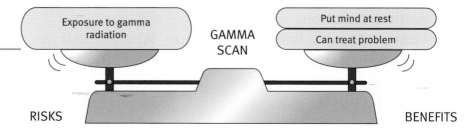

Jo's mum weighed the risk against the benefit and felt the investigation was worth it.

Jo's mum said 'We felt the risk was very small. It was worth it to find out what was wrong. Even with ordinary medicines, there can be risks. You have to weigh these things up. Nothing is completely safe.'

Treatment for thyroid cancer

Alf has thyroid cancer. First, he will have surgery to remove the tumour. Then he must have **radiotherapy**, to kill any cancer cells that may remain.

A hospital leaflet describes what will happen.

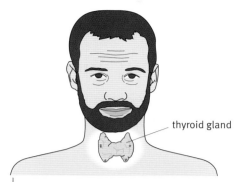

The thyroid gland is located in the front of the neck, below the voice box.

Radioiodine treatment

You will have to come in to hospital for a few days. You will stay in a single room.

You will be given a capsule to swallow, which contains iodine-131. This form of iodine is radioactive. You cannot eat or drink anything else for a couple of hours.

- The radioiodine is absorbed in your body.
- Radioiodine naturally collects in your thyroid, because this gland uses iodine to make its hormone.
- The radioiodine gives out beta radiation, which is absorbed in the thyroid.
- Any remaining cancer cells should be killed by the radiation.

You will have to stay in your room and take some precautions for the safety of visitors and staff. You will remain in hospital for a few days, until the amount of radioactivity in your body has fallen sufficiently.

Summary box
- Radioactive materials can be used to diagnose and treat medical problems.
- There are benefits and risks when using radioactive materials.

Questions

1. Look at the precautions that Jo has to take after the scan. Write a few sentences explaining to Jo why she has to do each of them.
2. It would be safe to stand next to Jo but not to kiss her. Use the words 'irradiation' and 'contamination' to explain why.
3. What are the risks and the benefits to Jo of having the treatment?
4. Suggest how the risk to Alf's family and other patients is kept as low as possible.
5. Explain why a half-life of eight days is more suitable than:
   a eight minutes   b eight years

Nuclear power

Find out about

- energy from nuclear fission
- nuclear power stations

Nuclear fission

Radioactive atoms have an unstable nucleus. Some nuclei can be made so unstable that they split in two. This process is called **nuclear fission**.

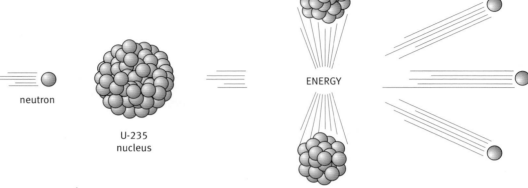

Splitting the nucleus of an atom.

When the nucleus breaks apart a small amount of the mass of the nucleus is converted to a huge amount of energy.

So the products of nuclear fission have a lot of kinetic energy. Each fission reaction produces roughly a million times more energy than when a molecule changes during a chemical reaction.

The devastating power of a nuclear weapon.

Nuclear weapons

During the Second World War there was a race to 'split the atom' and use the energy in a bomb.

On 16 July 1945, in a desert in the USA, a group of scientists tested 'the gadget'. Some thought it would not work. Others worried that it might destroy the atmosphere.

At 5.29 a.m., it was detonated and filled the skies with light. The bomb vaporized the metal tower supporting it. All desert sand within a distance of 700 m was turned into glass.

Some of the scientists were worried about the power of the bomb and wanted the project stopped. A few weeks later, the Americans dropped two nuclear bombs on Japan.

Generating electricity

A nuclear power station uses a nuclear reactor. This is designed to release the energy of at a slow and steady rate.

The fission takes place in the nuclear fuel. This makes them extremely hot.

A fluid, called a coolant, is pumped through the reactor. The hot fuel rods heat the coolant to around 500 °C. It then flows through a heat exchanger in the boiler, turning water into steam.

The steam is used in the same way as in a coal- or gas-fired power station. One reason for building nuclear power stations is to reduce the need for fossil fuels.

Nuclear weapons

Countries sometimes build nuclear reactors to make nuclear material for weapons. Nuclear weapons inspectors try to ensure that nuclear power stations are very secure, account for all their waste and are not operated in unstable countries.

The reactor core is sealed and shielded. Very little radiation gets out.

When the fuel rods are being used they become more radioactive because of the waste products.

Questions

1. Explain why these countries might decide to build, or not to build, nuclear power stations.
   a. No reserves of gas or coal.
   b. A neighbouring country that has nuclear weapons.
   c. A small land area, and a large population.
   d. A population worried about climate change.
   e. A history of nuclear accidents.
2. Suggest how gamma radiation from a nuclear reactor is contained, so that living things are not irradiated.
3. Write down two risks and two benefits of living in a country with nuclear power stations.

Summary box

- In nuclear fission a nucleus splits, releasing energy. The energy released is much greater than in a chemical reaction.

J Nuclear waste

Find out about

- the UK's nuclear waste
- the half-life of radioactive materials
- possible methods of disposal

Nuclear waste in the UK

The Nuclear Decommissioning Agency (NDA) is responsible for cleaning up nuclear waste. Most of the radioactive waste comes from power stations. The rest comes from medical uses, industry and scientific research. In addition to this 'everyday' waste, when power stations are too old to be used anymore the waste radioactive materials must be taken away to be stored. The waste is called the UK's 'nuclear legacy'.

A long-term hazard

Radioactive waste has very little effect on the UK's average background radiation. But it is still hazardous. This is because of contamination. Imagine that some waste leaks into the water supply. This could be taken up by food, which you eat. The radioactive material is now in your stomach, where it can irradiate your internal organs.

Some radioactive materials last for thousands of years. They must be kept safe and secure for all that time.

High-level radioactive waste is hot, so it is stored underwater.

Types of waste

The nuclear industry deals with three types of nuclear waste.

- **High-level waste** (HLW). This is 'spent' fuel rods. HLW gets hot because it is so radioactive. It has to be stored carefully but it doesn't last long. And there isn't very much of it. All the UK's HLW is kept in a pool of water at Sellafield.

- **Intermediate-level waste** (ILW). This is less radioactive than HLW. But the amount of ILW is increasing, as HLW decays to become ILW.

- **Low-level waste** (LLW). Protective clothing and medical equipment can be slightly radioactive. It is packed in drums and dumped in a landfill site that has been lined to prevent leaks.

The control room at a nuclear waste storage plant. People monitor the waste continuously.

Type of waste	Volume (m³)	Radioactivity
LLW	196 000	weak
ILW	92 500	strong
HLW	1 730	extremely strong

The amount of nuclear waste in store (2007). The problem of what to do with it remains unsolved.

Sellafield

Sellafield, in Cumbria, is the biggest nuclear site in the UK. Thousands of people work there. Sellafield processes nuclear waste and stores it ready for permanent disposal.

Keeping risks low is very important at Sellafield. They have plans to maintain production and safety if anything goes wrong.

Intermediate-level waste is the biggest technical challenge, because it is very long-lived. Currently it is chopped up, mixed with concrete, and stored in thousands of large stainless-steel containers. This is secure but not permanent. The long-term solution has to be secure and permanent.

The work of the NDA

Managing waste is very expensive. In 2010 the NDA spent £28 billion. There have been a number of public consultations about what to do with the waste. At the time of writing these are still going on. The preferred plan at the moment is to store it until a safe site can be found to bury it.

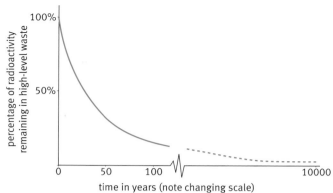

High Level Waste decays quickly at first. When its activity falls, it becomes Intermediate Level Waste. ILW stays radioactive for thousands of years.

When will it be 'safe'?

We are exposed to some radiation all the time (background radiation). When the nuclear waste only emits very low levels of radiation, similar to the background radiation, it poses little risk. The longer the half-life of the radioactive material, the longer it will take to become 'safe'.

Questions

1 Explain why disposing of ILW needs to be both
  a secure and b permanent

2 What are the advantages and disadvantages of keeping all the waste together above ground rather than burying it in a deep shaft and sealing it?

3 A small amount of nuclear fuel produces a lot of energy, so in the 1950s scientists thought this would be a cheap way of generating electricity. Explain why the real cost is much greater than realised at the time.

Summary box
- Nuclear waste is radioactive and must be safely stored for thousands of years.

Nuclear fusion

Find out about
- nuclear fusion
- the iter project

In nuclear fusion the nuclei of two small atoms join together and energy is released.

The diagram below shows one possible hydrogen fusion reaction. Two hydrogen nuclei fuse to make a helium nucleus, and there is a neutron left over.

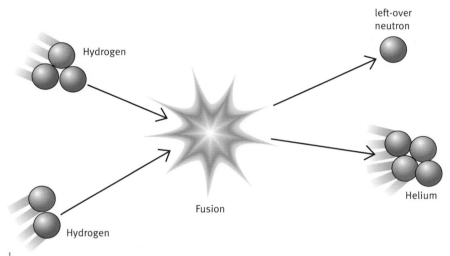

Fusion of two hydrogen nuclei gives helium.

The two hydrogen nuclei in the diagram have different numbers of neutrons, but you can tell they are both hydrogen because they have just one proton in the nucleus.

Helium is formed. It has two protons. Helium is an unreactive gas.

Positive charges repel

It is difficult to get the two hydrogen nuclei close enough to fuse, because both nuclei have a positive electric charge. Similar charges repel. The hydrogen nuclei must have enough energy to overcome this force and collide, so that they can fuse.

When the nuclei do fuse the new nucleus is more stable, and energy is released. The amount of energy released is very large.

The quest for fusion power stations

This is what scientists want to do:
- fuse the hydrogen nuclei from water to give helium nuclei
- use the energy released by this fusion reaction to generate electricity in power stations.

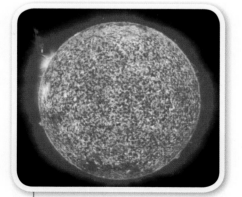

Hydrogen is fused to helium in the Sun.

The advantages are:
- the fuel is water, so there is plenty and it is cheap
- nuclear fusion does not produce as much radioactive waste as nuclear fission.

Over the past 70 years there has been a lot of research. Scientists can give hydrogen nuclei enough energy to overcome the repulsive force. The problem is controlling the reaction and keeping it going.

When hydrogen is heated to a very high temperature, the atoms lose their electrons and form a cloud of charged particles called a **plasma**. This is kept from touching the sides of the container by using magnetic fields. The JET project in the UK has researched fusion for many years. So far no reactor has produced more energy than it used.

The ITER project

This is a joint project between China, the European Atomic Energy Community, India, Japan, Korea, Russia and the USA. ITER means 'the way' in Latin. Fusion research is very expensive so these countries have joined together to build a research reactor in France. Construction has begun. It will take 10 years to build, and be used for research for 20 years. ITER will investigate how plasmas behave during the hydrogen fusion reaction at 150 million °C. One day they hope to build a nuclear fusion power station.

The H bomb

Hydrogen bombs, which fuse hydrogen, release hundreds of times more energy than atomic (fission) bombs. They are triggered using an atomic bomb to compress the hydrogen so that it fuses.

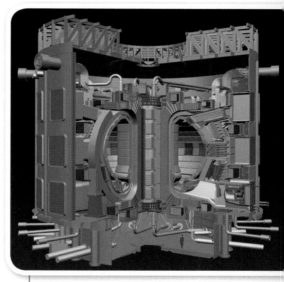

The planned ITER fusion reactor. Fusion will take place in the doughnut-shaped hole.

Questions

1. When two hydrogen nuclei fuse what element is formed?
2. a What is the electric charge on a hydrogen nucleus?
   b Explain whether two hydrogen nuclei attract or repel.
3. What are the advantages of countries working together on the ITER project?
4. ITER is very expensive. Write a letter to persuade the Government EITHER:
   a to stay part of ITER or
   b to leave ITER to save money.

Summary box
- ✓ Nuclear fusion releases energy.
- ✓ Hydrogen nuclei fuse to form helium nuclei.

Science Explanations

Our understanding of radioactivity and the structure of the atom has enabled many applications, such as nuclear power stations and cancer treatment, to be developed. Knowledge of the way ionising radiation behaves is essential for working safely and making good risk assessments.

You should know:

- why some materials are radioactive and emit ionising radiation all the time
- how ionising radiation can damage living cells
- that atoms have a nucleus made of protons and neutrons and is surrounded by electrons
- about the alpha scattering experiment and how it showed that the atom has a small, massive, positively charged nucleus
- that there are alpha and beta particles and gamma radiation
- about the different penetration properties of alpha, beta, and gamma radiations
- that there is background radiation all around us, mostly from natural sources
- what radiation dose measures, and what factors affect it
- the difference between contamination and irradiation
- how to interpret data on risk related to radiation dose
- that radioactive materials randomly emit ionising radiation all the time and that the rate of decay cannot be changed by physical or chemical changes
- that the activity of a radioactive source decreases over time
- what is meant by the half-life of a radioactive isotope
- that radioactive sources have a wide range of half-life values
- about how the half-life of a radioactive source affects the time it takes to become safe
- about uses of ionising radiation from radioactive materials and the people who work with them
- that nuclear fuels release energy when the nucleus changes during nuclear fission
- about the three categories of radioactive waste, and the different methods of disposal
- that hydrogen nuclei can fuse together to form helium if they are brought close enough together and this releases energy
- that the energy released in a nuclear reaction is much greater than that released in a chemical reaction such as burning a similar mass of fuel.

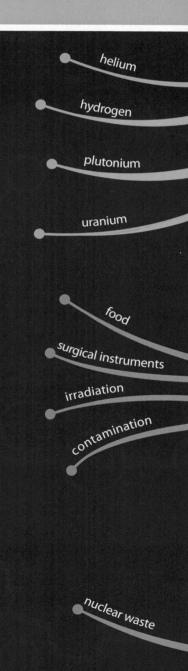

helium
hydrogen
plutonium
uranium
food
surgical instruments
irradiation
contamination
nuclear waste

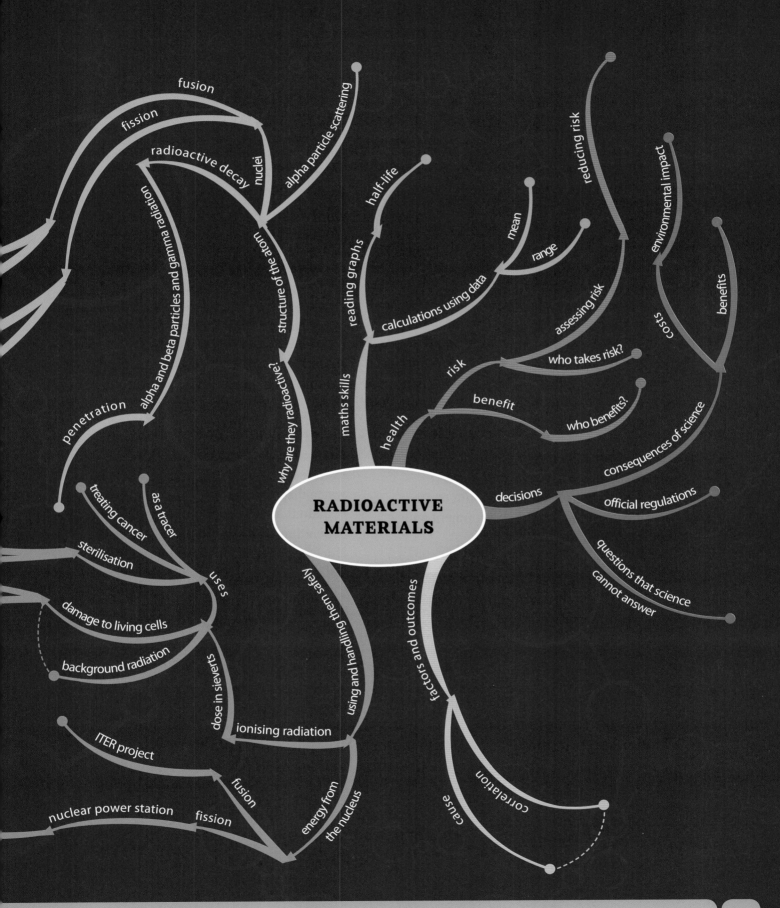

Ideas about Science

In addition to developing an understanding of radioactive materials, it is important to appreciate the risks involved and how we make decisions about using science and technology. When considering risk, you should be able to:

- explain that nothing is completely safe. Everything we do has a certain risk. Background radiation is all around us, so there is always a risk of our cells being harmed by ionising radiation. But if the dose is low the risk is very small.
- list some of the uses of radioactive materials and the risks arising from them, both to people working with radioactive sources and to the environment.
- describe some of the ways that we reduce these risks.
- use data to compare and discuss different risks. Compare the risks of living in a high-radon area with risks of other activities, for example, smoking tobacco.
- discuss decisions involving risk. To decide whether to build a type of nuclear power station you would need to take account of the chance of contaminating the environment and how serious that would be.
- identify risks and benefits to individuals and groups. Many medical treatments make use of radioactive isotopes. The risk and the benefit to the patient must be weighed up.
- take into account, in making decisions, who benefits and who takes the risks, for example, when deciding whether to build a nuclear power station.
- suggest benefits of activities known to have risk, for example, a scan that involves injecting a radioactive tracer into the body.
- suggest reasons why people are willing (or reluctant) to take a risk. For example, people who are ill may choose to have treatment in the hope that they will be cured. Some people may refuse treatment if doctors say they must have it, as they prefer to choose for themselves.

In making decisions about science and technology, you should be able to:

- identify the groups affected by a decision, and the main benefits and costs for each group, for example, when deciding on a location for a waste-disposal site.
- explain that different decisions may be made depending on social and economic factors. Nuclear power stations are built in remote areas where they will affect fewer people. Countries with no other resources for generating electricity may choose to build nuclear power stations.
- explain that there is official regulation of some areas of research and application of knowledge. Countries that have nuclear power stations keep account of all the radioactive waste, as this could be processed to produce nuclear weapons.
- distinguish questions that can be answered using a scientific approach from those that cannot, for example, 'Shall we build a nuclear power station at this site?' If there is no scientific reason why not, the final decision still depends on what society wants to do.

Review Questions

1. Copy and complete the sentences about radioactivity. Use key words from the module.

   _____ radiation is produced by radioactive _____ . The radiation is produced when an _____ nucleus _____ . The three types of radiation produced are: _____ particles, _____ particles, and _____ , which is electromagnetic radiation.

2. Rutherford had to develop a new model of the atom to explain the results of the scattering experiment.
   a How did he explain the fact that a few alpha particles were scattered back towards the source?
   b Most alpha particles passed straight through the foil. How did Rutherford explain this?

3. An isotope has a half-life of 74 days. Its activity is measured at 10 000 decays per second. It emits alpha radiation.
   a What does a half-life of 74 days mean?
   b Why could the isotope not be used to work out the age of rocks?

4. What is the difference between nuclear fission and nuclear fusion?

5. The table shows some of the radioactive isotopes that are used in a range of applications.

Isotope	Radiation emitted	Half-life
americium-241	alpha	430 years
carbon-14	beta	5700 years
cobalt-60	gamma	5 years
iodine-123	gamma	13 hours
iodine-131	beta	8 days
strontium-90	beta	29 years
uranium-235	alpha	700 million years

   For each application listed below choose the isotope you think most suitable and justify your answer, referring to both the half-life and the radiation emitted.
   a calculating the age of rocks
   b dating an ancient leather belt
   c monitoring the thickness of paper in a factory
   d monitoring uptake of iodine by the thyroid
   e detecting smoke
   f sterilising medical equipment

6. Nuclear power stations use radioactive materials to generate electricity. The waste from this process is radioactive. Explain how the three different categories of nuclear waste are treated.

7. Describe four sources of background radiation and explain whether each one is natural or man made.

Glossary

abundant Abundance measures how common an element is. Silicon is abundant in the lithosphere. Nitrogen is abundant in the atmosphere.

acceleration The rate of change of an object's velocity, that is, its change of velocity per second. In situations where the direction of motion is not important, the change of speed per second tells you the acceleration.

acid A compound that dissolves in water to give a solution with a pH lower than 7. Acid solutions change the colour of indicators, form salts when they neutralize alkalis, react with carbonates to form carbon dioxide, and give off hydrogen when they react with a metal. An acid is a compound that contains hydrogen in its formula and produces hydrogen ions when it dissolves in water.

action at a distance An interaction between two objects that are not in contact, where each exerts a force on the other. Examples include two magnets, two electric charges, or two masses (for example, the Earth and the Moon)

active site The part of an enzyme that the reacting molecules fit into.

active transport Molecules are moved in or out of a cell using energy. This process is used when transport needs to be faster than diffusion, and when molecules are being moved from a region where they are at low concentration to where they are at high concentration.

activity The rate at which nuclei in a sample of radioactive material decay and give out alpha, beta, or gamma radiation.

actual yield The mass of the required chemical obtained after separating and purifying the product of a chemical reaction.

aerobic respiration Respiration that uses oxygen.

air resistance The force exerted on an object by the air, when it moves through it. Its direction is opposite to the direction in which the object is moving.

alkali A compound that dissolves in water to give a solution with a pH higher than 7. An alkali can be neutralized by an acid to form a salt. Solutions of alkalis contain hydroxide ions.

alkali metal An element in Group 1 of the periodic table. Alkali metals react with water to form alkaline solutions of the metal hydroxide.

alloy A mixture of metals. Alloys are often more useful than pure metals.

alpha radiation The least penetrating type of ionising radiation, produced by the nucleus of an atom in radioactive decay. A high-speed helium nucleus.

alternating current (a.c.) An electric current that reverses direction many times a second.

Alzheimer's disease A form of senile dementia caused by irreversible degeneration of the brain.

amino acids The small molecules that are joined in long chains to make proteins. All the proteins in living things are made from 20 different amino acids joined in different orders.

ammeter A meter that measures the size of an electric current in a circuit.

ampere (amp) The unit of electric current.

anaerobic respiration Respiration that does not use oxygen.

antibodies A group of proteins made by white blood cells to fight dangerous microorganisms. A different antibody is needed to fight each different type of microorganism. Antibodies bind to the surface of the microorganism, which triggers other white blood cells to digest them.

atmosphere The layer of gases that surrounds the Earth.

attract Pull towards.

attractive forces (between molecules) Forces that try to pull molecules together. Attractions between molecules are weak. Molecular chemicals have low melting points and boiling points because the molecules are easy to separate.

auxin A plant hormone that affects plant growth and development. For example, auxin stimulates growth of roots in cuttings.

average speed The distance moved by an object divided by the time taken for this to happen.

axon A long, thin extension of the cytoplasm of a neuron. The axon carries electrical impulses very quickly.

background radiation The low-level radiation, mostly from natural sources, that everyone is exposed to all the time, everywhere.

bacterium (plural bacteria) One type of single-celled microorganism. They do not have a nucleus. Some bacteria may cause disease.

balanced equation An equation showing the formulae of the reactants and products. The equation is balanced when there is the same number of each kind of atom on both sides of the equation.

base pairing The bases in a DNA molecule (A, C, G, T) always bond in the same way. A and T always bond together. C and G always bond together.

behaviour Everything an organism does; its response to all the stimuli around it.

beta blockers Drugs that block the receptor sites for the hormone adrenaline. They inhibit the normal effects of adrenaline on the body.

beta radiation One of several types of ionising radiation, produced by the nucleus of an atom in radioactive decay. More penetrating than alpha radiation but less penetrating than gamma radiation. A high-speed electron.

bioethanol Ethanol fuel produced by yeast fermentation of plant materials, such as cane sugar and sugar beet.

biogas Methane gas produced by the anaerobic digestion of organic material, such as farm animal manure.

bleach A chemical that can destroy unwanted colours. Bleaches also kill bacteria. A common bleach is a solution of chlorine in sodium hydroxide.

bulk chemicals Chemicals made by industry on a scale of thousands or millions of tonnes per year. Examples are sulfuric acid, nitric acid, sodium hydroxide, ethanol, and ethanoic acid.

burette A graduated tube with taps or valves used to measure the volume of liquids or solutions during quantitative investigations such as titrations.

carbohydrate A natural chemical made of carbon, hydrogen, and oxygen. An example is glucose $C_6H_{12}O_6$. Carbohydrates include sugars, starch, and cellulose.

carbonate A compound that contains carbonate ions, CO_3^{2-}. An example is calcium carbonate, $CaCO_3$.

cartilage Tough, flexible tissue found at the end of bones and in joints. It protects the end of bones from rubbing together and becoming damaged.

catalyst A chemical that starts or speeds up a chemical reaction but is not used up in the process.

cell The basic structural and functional unit of all living things.

cell membrane Thin layer surrounding the cytoplasm of a cell. It restricts the passage of substances into and out of the cell.

cell wall Rigid outer layer of plant cells and bacteria.

cellulose The chemical that makes up most of the fibre in food. The human body cannot digest cellulose.

central nervous system In mammals the brain and spinal cord.

cerebral cortex The highly folded outer region of the brain, concerned with conscious behaviour.

chain reaction A process in which the products of one nuclear reaction cause further nuclear reactions to happen, so that more and more reactions occur and more and more product is formed. Depending on how this process is controlled, it can be used in nuclear weapons or the nuclear reactors in power stations.

charged Carrying an electric charge. Some objects (such as electrons and protons) are permanently charged. A plastic object can be charged by rubbing it. This transfers electrons to or from it.

chemical change/reaction A change that forms a new chemical.

chemical equation A summary of a chemical reaction showing the reactants and products with their physical states (see balanced chemical equation).

chemical industry The industry that converts raw materials such as crude oil, natural gas, and minerals into useful products such as pharmaceuticals, fertilisers, paints, and dyes.

chemical properties A chemical property describes how an element or compound interacts with other chemicals, for example, the reactivity of a metal with water.

chemical species The different chemical forms that an element can take, for example, chlorine has three chemical species: atom, molecule, and ion. Each of these forms has distinct properties.

chlorophyll A green pigment found in chloroplasts. Chlorophyll absorbs energy from sunlight for photosynthesis.

chloroplast An organelle found in some plant cells where photosynthesis takes place.

chromosome Long, thin, threadlike structures in the nucleus of a cell made from a molecule of DNA. Chromosomes carry the genes.

clone A new cell or individual made by asexual reproduction. A clone has the same genes as its parent.

collision theory The theory that reactions happen when molecules collide. The theory helps to explain the factors that affect the rates of chemical change. Not all collisions between molecules lead to reaction.

commutator An device for changing the direction of the electric current through the coil of a motor every half turn. It consists of a ring divided into two halves (a split ring) with two contacts (brushes) touching the two halves.

concentrated solution The concentration of a solution depends on how much dissolved chemical (solute) there is compared with the solvent. A concentrated solution contains a high level of solute to solvent.

concentration The quantity of a chemical dissolved in a stated volume of solution. Concentrations can be measured in grams per litre.

conditioned reflex A reflex where the response is associated with a secondary stimulus, for example, a dog salivates when it hears a bell because it has associated the bell with food.

conditioning Reinforcement of behaviour associated with conditioned reflexes.

conscious To have awareness of surroundings and sensations.

consciousness The part of the human brain concerned with thought and decision making.

conservation of energy The fundamental idea that the total amount of energy in the universe is constant, and never increases or decreases. So if something loses energy, one or more other things must have gained the same amount of energy.

contamination (radioactive) Having a radioactive material inside the body, or having it on the skin or clothes.

control rod In a nuclear reactor, rods made of a special material that absorbs neutrons are raised and lowered to control the rate of fission reactions.

coolant In a nuclear reactor, the liquid or gas that circulates through the core and transfers heat to the boiler.

corrosive A corrosive chemical may destroy living tissue on contact.

counter-force A force in the opposite direction to something's motion.

covalent bonding Strong attractive forces that hold atoms together in molecules. Covalent bonds form between atoms of non-metallic elements.

crust (of the Earth) The outer layer of the lithosphere.

crystalline A material with molecules, atoms, or ions lined up in a regular way as in a crystal.

cutting A shoot or leaf taken from a plant, to be grown into a new plant.

cytoplasm Gel enclosed by the cell membrane that contains the cell organelles such as mitochondria.

decommissioning Taking a power station out of service at the end of its lifetime, dismantling it, and disposing of the waste safely.

denatured A change in the usual nature of something. When enzymes are denatured by heat, their structure, including the shape of the active site, is altered.

development How an organism changes as it grows and matures. As a zygote develops, it forms more and more cells. These are organised into different tissues and organs.

diamond A gemstone. A form of carbon. It has a giant covalent structure and is very hard.

diatomic A molecule with two atoms, for example, N_2, O_2, and Cl_2

diffusion Movement of molecules from a region of high concentration to a region of lower concentration.

dilute The concentration of a solution depends on how much dissolved chemical (solute) there is compared with the solvent. A dilute solution contains a low level of solute to solvent.

direct current (d.c.) An electric current that stays in the same direction.

displacement reaction A more reactive halogen will displace a less reactive halogen, for example, chlorine will displace bromide ions to form bromine and chloride ions.

dissolve Some chemicals dissolve in liquids (solvents). Salt and sugar, for example, dissolve in water.

distance The length of the path along which an object moves.

distance–time graph A useful way of summarising the motion of an object by showing how far it has moved from its starting point at every instant during its journey.

distance–time graph A graph showing the distance an object moves along its path at each moment during its journey.

double helix The shape of the DNA molecule, with two strands twisted together in a spiral.

driving force The force pushing something forward, for example, a bicycle.

Ecstasy A recreational drug that increases the concentration of serotonin at the synapses in the brain, giving pleasurable feelings. Long-term effects may include destruction of the synapses.

effector The part of a control system that brings about a change to the system.

electric charge A fundamental property of matter. Electrons and protons are charged particles. Objects become charged when electrons are transferred to or from them, for example, by rubbing.

electric circuit A closed loop of conductors connected between the positive and negative terminals of a battery or power supply.

electric current A flow of charges around an electric circuit.

electric field A region where an electric charge experiences a force. There is an electric field around any electric charge.

electrode A conductor made of a metal or graphite through which a current enters or leaves a chemical during electrolysis. Electrons flow into the negative electrode (cathode) and out of the positive electrode (anode).

electrolysis Splitting up a chemical into its elements by passing an electric current through it.

electrolyte A chemical that can be split up by an electric current when molten or in solution is the electrolyte. Ionic compounds are electrolytes.

electromagnetic induction The name of the process in which a potential difference (and hence often an electric current) is generated in a wire, when it is in a changing magnetic field.

electron A tiny, negatively charged particle, which is part of an atom. Electrons are found outside the nucleus. Electrons have negligible mass and one unit of charge.

electron arrangement The number and arrangement of electrons in an atom of an element.

electrostatic attraction The force of attraction between objects with opposite electric charges.

embryonic stem cell Unspecialised cell in the very early embryo that can divide to form any type of cell, or even a whole new individual. In human embryos the cells are identical and unspecialised up to the eight-cell stage.

end point The point during a titration at which the reaction is just complete. For example, in an acid–alkali titration, the end point is reached when the indicator changes colour. This happens when exactly the right amount of acid has been added to react with all the alkali present at the start.

endothermic An endothermic process takes in energy from its surroundings.

energy level The electrons in an atom have different energies and are arranged at distinct energy levels.

energy-level diagram A diagram to show the difference in energy between the reactants and the products of a reaction.

enzyme A protein that catalyses (speeds up) chemical reactions in living things.

ethanol Waste product from anaerobic respiration in plants and yeast.

exothermic An exothermic process gives out energy to its surroundings.

extraction (of metals) The process of obtaining a metal from a mineral by chemical reduction or electrolysis. It is often necessary to concentrate the ore before extracting the metal.

fatty sheath Fat wrapped around the outside of an axon to insulate neurons from each other.

feral Untamed, wild.

fermentation Chemical reactions in living organisms that release energy from organic chemicals, such as yeast producing alcohol from the sugar in grapes.

fermenter A large vessel in which microorganisms are grown to make a useful product.

fetus A developing human embryo is referred to as a fetus once it reaches eight weeks after fertilization. A fetus already has all the main organs that it will have at birth.

fine chemicals Chemicals made by industry in smaller quantities than bulk chemicals. Fine chemicals are used in products such as food additives, medicines, and pesticides.

flame colour A colour produced when a chemical is held in a flame. Some elements and their compounds give characteristic colours. Sodium and sodium compounds, for example, give bright yellow flames.

food chain In the food industry this covers all the stages from where food grows, through harvesting, processing, preservation, and cooking to being eaten.

force A push or a pull experienced by an object when it interacts with another. A force is needed to change the motion of an object.

formulae (chemical) A way of describing a chemical that uses symbols for atoms. A formula gives information about the numbers of different types of atom in the chemical. The formula of sulfuric acid, for example, is H_2SO_4.

friction The force exerted on an object due to the interaction between it and another object that it is sliding over. It is caused by the roughness of both surfaces at a microscopic level.

fuel rod A container for nuclear fuel, which enables fuel to be inserted into, and removed from, a nuclear reactor while it is operating.

gametes The sex cells that fuse to form a zygote. In humans, the male gamete is the sperm and the female gamete is the egg.

gamma radiation (gamma rays) The most penetrating type of ionising radiation, produced by the nucleus of an atom in radioactive decay. The most energetic part of the electromagnetic spectrum.

gas exchange The exchange of oxygen and carbon dioxide that takes place in the lungs.

gene A section of DNA giving the instructions for a cell about how to make one kind of protein.

gene switching Genes in the nucleus of a cell switch off and are inactive when a cell becomes specialised. Only genes that the cell needs to carry out its particular job stay active.

generator A device that uses motion to generate electricity. It consists of a coil that rotates in a magnetic field. This produces a potential difference across the ends of the coil, which can then be used to provide an electric current.

genetic Factors that are affected by an organism's genes.

genetic variation Differences between individuals caused by differences in their genes. Gametes show genetic variation – they all have different genes.

giant covalent structure A giant, three-dimensional arrangement of atoms that are held together by covalent bonds. Silicon dioxide and diamond have giant covalent structures.

giant ionic lattice The structure of solid ionic compounds. There are no individual molecules, but millions of oppositely charged ions packed closely together in a regular, three-dimensional arrangement.

glands Parts of the body that make enzymes, hormones, and other secretions in the body, for example, sweat glands.

glucose Sugar produced during photosynthesis.

graphite A form of carbon. It has a giant covalent structure. It is unusual for a non-metal in that it conducts electricity.

gravitational potential energy The energy stored when an object is raised to a higher point in the Earth's gravitational field.

group Each column in the periodic table is a group of similar elements.

habitat The place where an organism lives.

haemoglobin The protein molecule in red blood cells. Haemoglobin binds to oxygen and carries it around the body. It also gives blood its red colour.

half-life The time taken for the amount of a radioactive element in a sample to fall to half its original value.

halogens The family name of the Group 7 elements.

harmful A harmful chemical is one that may cause damage to health if swallowed, breathed in, or absorbed through the skin.

high-level waste A category of nuclear waste that is highly radioactive and hot. Produced in nuclear reactors and nuclear-weapons processing.

hormone A chemical messenger secreted by specialised cells in animals and plants. Hormones bring about changes in cells or tissues in different parts of the animal or plant.

hydrogen ion A hydrogen atom that has lost one electron. The symbol for a hydrogen ion is H+. Acids produce aqueous hydrogen ions, H+(aq), when dissolved in water.

hydrosphere All the water on Earth. This includes oceans, lakes, rivers, underground reservoirs, and rainwater.

hydroxide ion A negative ion, OH–. Alkalis give aqueous hydroxide ions when they dissolve in water.

in parallel A way of connecting electric components that makes a branch (or branches) in the circuit so that charges can flow around more than one loop.

in series A way of connecting electric components so that they are all in a single loop. The charges pass through them all in turn.

indicator A chemical that shows whether a solution is acidic or alkaline. For example, litmus turns blue in alkalis and red in acids. Universal indicator has a range of colours that show the pH of a solution.

innate Inborn, inherited from parents via genes.

insoluble Does not form a solution (dissolve) in water or other solutes.

instantaneous speed The speed of an object at a particular instant. In practice, its average speed over a very short time interval.

interaction What happens when two objects collide, or influence each other at a distance. When two objects interact, each experiences a force.

interaction pair Two forces that arise from the same interaction. They are equal in size and opposite in direction, and each acts on a different object.

intermediate-level waste A category of nuclear waste that is generally short-lived but requires some shielding to protect living organisms, for example contaminated materials that result from decommissioning a nuclear reactor.

involuntary An automatic response made by the body without conscious thought.

ion An electrically charged atom or group of atoms.

ionic bonding Very strong attractive forces that hold the ions together in an ionic compound. The forces come from the attraction between positively and negatively charged ions.

ionic compounds Compounds formed by the combination of a metal and a non-metal. They contain positively charged metal ions and negatively charged non-metal ions.

ionic equation An ionic equation describes a chemical change by showing only the reacting ions in solution.

ionising Able to remove electrons from atoms, producing ions.

ionising radiation Radiation with photons of sufficient energy to remove electrons from atoms in its path. Ionising radiation, such as ultraviolet, X-rays, and gamma rays, can damage living cells.

irradiation Being exposed to radiation from an external source.

isotope Atoms of the same element that have different mass numbers because they have difference numbers of neutrons in the nucleus.

kinetic energy The energy that something has owing to its motion.

lactic acid Waste product from anaerobic respiration in animals.

learn To gain new knowledge or skills.

life cycle The stages an organism goes through as it matures, develops, and reproduces.

light intensity The amount of light reaching a given area.

light meter Device for measuring light intensity.

light-dependent resistor (LDR) An electric circuit component whose resistance varies depending on the brightness of light falling on it.

limiting factor The factor that prevents the rate of photosynthesis from increasing at a particular time. This may be light intensity, temperature, carbon dioxide concentration, or water availability.

line spectrum A spectrum made up of a series of lines. Each element has its own characteristic line spectrum.

lithosphere The rigid outer layer of the Earth, made up of the crust and the part of mantle just below it.

lock-and-key model In chemical reactions catalysed by enzymes, molecules taking part in the reaction fit exactly into the enzyme's active site. The active site will not fit other molecules – it is specific. This is like a key fitting into a lock.

long-term memory The part of the memory that stores information for a long period, or permanently.

low-level waste A category of nuclear waste that contains small amounts of short-lived radioactivity, for example, paper, rags, tools, clothing, and filters from hospitals and industry.

magnetic field The region around a magnet, or a wire carrying an electric current, in which magnetic effects can be detected. For example, another small magnet in this region will experience a force and may tend to move.

mantle The layer of rock between the crust and the outer core of the Earth. It is approximately 2900 km thick.

meiosis Cell division that halves the number of chromosomes to produce gametes. The four new cells are genetically different from each other and from the parent cell.

memory The storage and retrieval of information by the brain.

meristem cells Unspecialised cells in plants that can develop into any kind of specialised cell.

metal Elements on the left side of the periodic table. Metals have characteristic properties: they are shiny when polished and they conduct electricity. Some metals react with acids to give salts and hydrogen. Metals are present as positive ions in salts.

metal hydroxide A compound consisting of metal positive ions and hydroxide ions. Examples are sodium hydroxide, NaOH, and magnesium hydroxide, $Mg(OH)_2$.

metal oxide A compound of a metal with oxygen.

metallic bonding Very strong attractive forces that hold metal atoms together in a solid metal. The metal atoms lose their outer electrons and form positive ions. The electrons drift freely around the lattice of positive metal ions and hold the ions together.

mineral A naturally occurring element or compound in the Earth's lithosphere.

mitochondrion (plural mitochondria) An organelle in animal and plant cells where respiration takes place.

mitosis Cell division that makes two new cells identical to each other and to the parent cell.

models of memory Explanations for how memory is structured in the brain.

molecular models Models to show the arrangement of atoms in molecules, and the bonds between the atoms.

molecule A group of atoms joined together. Most non-metals consist of molecules. Most compounds of non-metals with other non-metals are also molecular.

molten A chemical in the liquid state. A chemical is molten when the temperature is above is melting point but below its boiling point.

momentum (plural momenta) A property of any moving object. Equal to mass multiplied by velocity.

motor A device that uses an electric current to produce continuous motion.

motor neuron A neuron that carries nerve impulses from the brain or spinal cord to an effector.

mRNA Messenger RNA, a chemical involved in making proteins in cells. The mRNA molecule is similar to DNA but single stranded. It carries the genetic code from the DNA molecule out of the nucleus into the cytoplasm.

multi-store model One explanation for how the human memory works.

muscles Muscles move parts of the skeleton for movement. There is also muscle tissue in other parts of the body, for example, in the walls of arteries.

negative A label used to name one type of charge or one terminal of a battery. It is the opposite of positive.

negative ion An ion that has a negative charge (an anion).

nerve cell A cell in the nervous system that transmits electrical signals to allow communication within the body.

nerve impulses Electrical signals carried by neurons (nerve cells).

nervous system Tissues and organs that control the body's responses to stimuli. In a mammal it is made up of the central nervous system and peripheral nervous system.

neuron Nerve cell.

neuroscientist A scientist who studies how the brain and nerves function.

neutralization reaction A reaction in which an acid reacts with an alkali to form a salt. During neutralization reactions, the hydrogen ions in the acid solution react with hydroxide ions in the alkaline solution to make water molecules.

neutrons An uncharged particle found in the nucleus of atoms. The relative mass of a neutron is 1.

newborn reflexes Reflexes to particular stimuli that usually occur only for a short time in newborn babies.

nitrate ions An ion is an electrically charged atom or group of atoms. The nitrate ion has a negative charge, NO_3-.

non-ionising radiation Radiation with photons that do not have enough energy to ionize molecules.

nuclear fission The process in which a nucleus of uranium-235 breaks apart, releasing energy, when it absorbs a neutron.

nuclear fuel In a nuclear reactor, each uranium atom in a fuel rod undergoes fission and releases energy when hit by a neutron.

nuclear fusion The process in which two small nuclei combine to form a larger one, releasing energy. An example is hydrogen combining to form helium. This happens in stars, including the Sun.

nucleus (atom) The tiny central part of an atom (made up of protons and neutrons). Most of the mass of an atom is concentrated in its nucleus.

nucleus (cell) Organelle that contains the chromosomes cells of plants, animals, fungi, and some microorganisms.

ohm The unit of electrical resistance. Symbol Ω.

Ohm's law The result that the current, I, through a resistor, R, is proportional to the voltage, V, across the resistor, provided its temperature remains the same. Ohm's law does not apply to all conductors.

optimum temperature The temperature at which enzymes work fastest.

ore A natural mineral that contains enough valuable minerals to make it profitable to mine.

organelles The specialised parts of a cell, such as the nucleus and mitochondria. Chloroplasts are organelles that occur only in plant cells.

organs Parts of a plant or animal made up of different tissues.

osmosis The diffusion of water across a partially permeable membrane.

oxidation A reaction that adds oxygen to a chemical.

oxide A compound of an element with oxygen.

pancreas An organ in the body that produces some hormones and digestive enzymes. The hormone insulin is made here.

partially permeable membrane A membrane that acts as a barrier to some molecules but allows others to diffuse through freely.

pathway A series of connected neurones that allow nerve impulses to travel along a particular route very quickly.

percentage yield A measure of the efficiency of a chemical synthesis.

period In the context of chemistry, a row in the periodic table.

periodic In chemistry, a repeating pattern in the properties of elements. In the periodic table one pattern is that each period starts with metals on the left and ends with non-metals on the right.

peripheral nervous system The network of nerves connecting the central nervous system to the rest of the body.

pH scale A number scale that shows the acidity or alkalinity of a solution in water.

phloem A plant tissue that transports sugar throughout a plant.

photons Tiny 'packets' of electromagnetic radiation. All electromagnetic waves are emitted and absorbed as photons. The energy of a photon is proportional to the frequency of the radiation.

photosynthesis The process in green plants that uses energy from sunlight to convert carbon dioxide and water into the sugar glucose.

phototropism The bending of growing plant shoots towards the light.

physical properties Properties of elements and compounds such as melting point, density, and electrical conductivity. These are properties that do not involve one chemical turning into another.

pilot plant A small-scale chemical processing facility. A pilot plant is used to test processes before scaling up to full-scale production.

plant A chemical plant is an industrial facility used to manufacture chemicals.

plasma A collection of electrons and nuclei that can be formed when a gas has so much energy that its atoms are fully ionised.

polymer A material made up of very long molecules. The molecules are long chains of smaller molecules.

positive A label used to name one type of charge, or one terminal of a battery. It is the opposite of negative.

positive ion Ions that have a positive charge (cations).

potential difference (p.d.) The difference in potential energy (for each unit of charge flowing) between any two points in an electric circuit.

power In an electric circuit, the rate at which work is done by the battery or power supply on the components in a circuit. Power is equal to current × voltage.

precipitate An insoluble solid formed on mixing two solutions. Silver bromide forms as a precipitate on mixing solutions of silver nitrate and potassium bromide.

proportional Two variables are proportional if there is a constant ratio between them.

protein Chemicals in living things that are polymers made by joining together amino acids.

proton Tiny particle present in the nuclei of atoms. Protons are positively charged (+1).

proton number The number of protons in the nucleus of an atom (also called the atomic number). In an uncharged atom this also gives the number of electrons.

Prozac A brand name for an antidepressant drug. It increases the concentration of serotonin at the synapses in the brain.

pupil reflex The reaction of the muscles in the pupil to light. The pupil contracts in bright light and relaxes in dim light.

quadrat A square grid of a known area that is used to survey plants in a location. Quadrats come in different sizes up to 1 m2. The size of quadrat that is chosen depends on the size of the plants and also the area that needs to be surveyed.

radiation A flow of information and energy from a source. Light and infrared are examples. Radiation spreads out from its source, and may be absorbed or reflected by objects in its path. It may also go (be transmitted) through them.

radiation dose A measure, in millisieverts, of the possible harm done to your body, which takes into account both the amount and type of radiation you are exposed to.

radioactive Used to describe a material, atom, or element that produces alpha, beta, or gamma radiation.

radioactive dating Estimating the age of an object such as a rock by measuring its radioactivity. Activity falls with time, in a way that is well understood.

radioactive decay The spontaneous change in an unstable element, giving out alpha, beta, or gamma radiation. Alpha and beta emission result in a new element.

radiotherapy Using radiation to treat a patient.

random Of no predictable pattern.

rate of photosynthesis Rate at which green plants convert carbon dioxide and water to glucose in the presence of light.

rate of reaction A measure of how quickly a reaction happens. Rates can be measured by following the disappearance of a reactant or the formation of a product.

reactants The chemicals on the left-hand side of an equation. These chemicals react to form the products.

reacting mass The masses of chemicals that react together, and the masses of products that are formed. Reacting masses are calculated from the balanced symbol equation using relative atomic masses and relative formula masses.

reaction (of a surface) The force exerted by a hard surface on an object that presses on it.

reactive metal A metal with a strong tendency to react with chemicals such as oxygen, water, and acids. The more reactive a metal, the more strongly it joins with other elements such as oxygen. So reactive metals are hard to extract from their ores.

receptor The part of a control system that detects changes in the system and passes this information to the processing centre.

receptor molecule A protein (often embedded in a cell membrane) that exactly fits with a specific molecule, bringing about a reaction in the cell.

recycling A range of methods for making new materials from materials that have already been used.

red blood cells Blood cells containing haemoglobin, which binds to oxygen so that it can be carried around the body by the bloodstream.

reducing agent A chemical that removes oxygen from another chemical. For example, carbon acts as a reducing agent when it removes oxygen from a metal oxide. The carbon is oxidized to carbon monoxide during this process.

reduction A reaction that removes oxygen from a chemical.

reflex arc A neuron pathway that brings about a reflex response. A reflex arc involves a sensory neuron, connecting neurons in the brain or spinal cord, and a motor neuron.

relative atomic mass The mass of an atom of an element compared to the mass of an atom of carbon. The relative atomic mass of carbon is defined as 12.

relative formula mass The combined relative atomic masses of all the atoms in a formula. To find the relative formula mass of a chemical, you just add up the relative atomic masses of the atoms in the formula.

relay neuron A neuron that carries the impulses from the sensory neuron to the motor neuron.

repel Push apart.

repetition Act of repeating.

repetition of information Saying or writing the same thing several times.

resistance The resistance of a component in an electric circuit indicates how easy or difficult it is to move charges through it.

respiration A series of chemical reactions in cells that release energy for the cell to use.

response Action or behaviour that is caused by a stimulus.

resultant force The sum, taking their directions into account, of all the forces acting on an object.

retina Light-sensitive layer at the back of the eye. The retina detects light by converting light into nerve impulses.

retrieval of information Collecting information from a particular source.

ribosomes Organelles in cells. Amino acids are joined together to form proteins in the ribosomes.

risk The probability of an outcome that is seen as undesirable, associated with some behaviour or process.

risk assessment A check on the hazards involved in a scientific procedure. A full assessment includes the steps to be taken to avoid or reduce the risks from the hazards identified.

rock A naturally occurring solid, made up of one or more minerals.

root hair cell Microscopic cell that increases the surface area for absorption of minerals and water by plant roots.

rooting powder A product used in gardening containing plant hormones. Rooting powder encourages a cutting to form roots.

salt An ionic compound formed when an acid neutralizes an alkali or when a metal reacts with a non-metal.

sample Small part of something that is likely to represent the whole.

scale up To redesign a synthesis to produce a chemical in larger amounts. A process might be scaled up first from a laboratory method to a pilot plant, then from a pilot plant to a full-scale industrial process.

sensory neuron A neuron that carries nerve impulses from a receptor to the brain or spinal cord.

serotonin A chemical released at one type of synapse in the brain, resulting in feelings of pleasure.

shell A region in space (around the nucleus of an atom) where there can be electrons.

short-term memory The part of the memory that stores information for a short time.

simple reflex An automatic response made by an animal to a stimulus.

slope The slope of a graph is a measure of its steepness.

small molecules Particles of chemicals that consist of small numbers of atoms bonded together. Chemicals made up of one or more non-metallic elements and that have low boiling and melting points consist of small molecules.

social behaviour Behaviour that takes place between members of the same species, including humans.

specialised A specialised cell is adapted for a particular job.

spectroscopy The use of instruments to produce and analyse spectra. Chemists use spectroscopy to study the composition, structure, and bonding of elements and compounds.

starch A type of carbohydrate found in bread, potatoes, and rice. Plants produce starch to store the energy food they make by photosynthesis. Starch molecules are a long chain of glucose molecules.

starch grains Microscopic granules of starch forming an energy store in plant cells.

static electricity Electric charge that is not moving around a circuit but has built up on an object such as a comb or a rubbed balloon.

stem cell Unspecialised animal cell that can divide and develop into specialised cells.

sterilisation The process of making something free from live bacteria and other microorganisms.

stimulus A change in the environment that causes a response.

stomata Tiny holes in the underside of a leaf that allow carbon dioxide into the leaf and water and oxygen out of the leaf.

strong (nuclear) force A fundamental force of nature that acts inside atomic nuclei.

structural proteins Proteins that are used to build cells.

subatomic particles The particles that make up atoms. Protons, neutrons, and electrons are subatomic particles.

surface area (of a solid chemical) The area of a solid in contact with other reactants that are liquids or gases.

sustainable Able to continue over long periods of time.

synapse A tiny gap between neurons that transmits nerve impulses from one neuron to another by means of chemicals diffusing across the gap.

tarnish When the surface of a metal becomes dull or discoloured because it has reacted with the oxygen in the air.

theoretical yield The amount of product that would be obtained in a reaction if all the reactants were converted to products exactly as described by the balanced chemical equation.

therapeutic cloning Growing new tissues and organs from cloning embryonic stem cells. The new tissues and organs are used to treat people who are ill or injured.

thermistor An electric circuit component whose resistance changes markedly with its temperature. It can therefore be used to measure temperature.

tissue Group of specialised cells of the same type working together to do a particular job.

tissue fluid Plasma that is forced out of the blood as it passes through a capillary network. Tissue fluid carries dissolved chemicals from the blood to cells.

titration An analytical technique used to find the exact volumes of solutions that react with each other.

toxic A chemical that may lead to serious health risks, or even death, if breathed in, swallowed, or taken in through the skin.

transect A straight line that runs through a location. Data on plant and animal distribution is recorded at regular intervals along the line.

transformer An electrical device consisting of two coils of wire wound on an iron core. An alternating current in one coil causes an ever-changing magnetic field that induces an alternating current in the other. Used to 'step' voltage up or down to the level required.

transmitter substance Chemical that bridges the gap between two neurons.

trend A description of the way a property increases or decreases along a series of elements or compounds, which is often applied to the elements (or their compounds) in a group or period.

triplet code A sequence of three bases coding for a particular amino acid in the genetic code.

unspecialised Cells that have not yet developed into one particular type of cell.

unstable The nucleus in radioactive isotopes is not stable. It is liable to change, emitting one of several types of radiation. If it emits alpha or beta radiation, a new element is formed.

velocity The speed of an object in a given direction. Unlike speed, which only has a size, velocity also has a direction.

velocity–time graph A useful way of summarising the motion of an object by showing its velocity at every instant during its journey.

voltage The voltage marked on a battery or power supply is a measure of the 'push' it exerts on charges in an electric circuit. The 'voltage' between two points in a circuit means the 'potential difference' between these points.

voltmeter An instrument for measuring the potential difference between two points in an electric circuit.

work Work is done whenever a force makes something move. The amount of work is force multiplied by distance moved in the direction of the force. This is equal to the amount of energy transferred.

working memory The system in the brain responsible for holding and manipulating information needed to carry out tasks.

xylem Plant tissue that transports water through a plant.

yeast Single-celled fungus used in brewing and baking.

yield The crop yield is the amount of crop that can be grown per area of land.

zygote The cell made when a sperm cell fertilises an egg cell in sexual reproduction.

Index

abundant elements 146
acceleration 85, 88
acids 228–233
action-at-a-distance forces 77
active sites 18
active working memory 218, 219
activity (radioactive decay) 270
adaptations 32, 33
adenine (A) 120, 121
adult stem cells 129
aerobic respiration 34
air bags 92
air resistance 83
alcohol 37, 38
alkali metals 48, 49
alkalis 228–230, 232
alpha radiation 257–259, 261, 262, 264, 265, 268
alternating current (a.c.) 187
aluminium 146, 154, 155, 158, 159
Alzheimer's disease 214
amino acids 18, 28, 122
ammeters 171
amoeba 196
anaerobic respiration 36–39
animals 16, 34, 36, 208, 209, 211
antibodies 124
argon 138
asexual reproduction 117
atmosphere 136–139
atoms
    carbon minerals 148, 149
    chemical species 68
    Earth 138, 139
    electric charge 168
    electrons 58, 59
    halogens 52
    ionic theory 66, 67
    masses 46, 153
    nuclear fission 274, 275
    nuclear fusion 278, 279
    radioactive materials 257–261, 270, 271
    structure 56, 57, 66, 67
ATP (adenosine triphosphate) molecules 35
attractive forces 138, 166, 167
auxins 111
average speed 84
axons 203–205

babies 108, 198, 199
background radiation 264
bacteria 16, 17, 20, 36, 37, 39, 263
balanced forces 82
ball-and-stick models 139
balloons 78
base pairing 120, 121
batteries 172–174, 178, 180, 181, 185
behaviour 196–198, 206, 208–211
beta radiation 257, 261, 262
bicycles 94, 95
biofuel 38, 39
blood cells 114
body temperature 19, 20
Bohr, Niels 58
boiling points 140
bonding 138, 139, 147, 156, 157
brain 194, 195, 206, 207
branching of electric circuits 172
bromide 145
bromine 52, 53, 61
building blocks 107
bulk chemicals 226
Bunsen burner 54
burettes 235

calcite 146
calcium carbonate 143
calcium chloride 244
calcium fluoride 52
cancer 129, 267, 269, 273
carbohydrates 22, 23, 27
carbon 137, 152
carbon dioxide (CO_2) 31, 37–39, 138–140
carbon minerals 148, 149
carbonates 143, 145, 231
cars 38, 79, 85, 91–93
catalysts 17–19, 240–242
cells 16, 23, 26, 107–111, 114–119, 124–129
cellulose 23
central nervous system (CNS) 202, 203
cerebral cortex 206
CFCs (chlorofluorocarbons) 68
change of momentum 90
charge 166–168, 278
chemicals 49–51, 134, 135, 224–227
    energy 236, 237
    patterns 44, 45
    species 68, 69

children 213
chloride 145, 147
chlorine 49, 52, 53, 61, 62, 64, 65, 68, 69
chlorophyll 22
chloroplasts 22, 26, 34
chromosomes 114–117, 119, 126
circuits, electric 169–183
clones 108, 111, 128
closed loop circuits 169
coils 184–187, 188
collision theory 242, 243
combs 166
commutators 185
compasses 184
complex behaviour 197, 206, 210
compounds, ionic theory 65
concentrated solutions 27
concentration 240, 241
conditioning 208, 209
conduction 156
consciousness 206
conservation of energy 99
contact forces 77
contamination 265
conventional current 170
coolants 275
copper 152, 157
copying of DNA 121
core of Earth 136
corrosive substances 52
covalent bonds/structures 138, 148, 149
crumple zones 91, 92
crust of Earth 136
crystals 49, 62, 64, 65, 146, 147, 149, 155
curare (toxin) 205
current 168, 169–181
cuttings 111
cytoplasm 26, 34, 37, 123
cytosine (C) 120, 121

Dalton, John 56
Darwin, Charles 113
daughter products 261
Davy, Humphry 64
decay (radioactive) 257, 261, 270
denaturation 20
development 104, 105
dialysis 244
diamond 148, 149, 257
diatomic molecules 52, 68
diffusion, plants 24–27, 30
dilute solutions 27

direct current (d.c.) 187
displacement reactions 53
dissolved compounds 140, 147
distance–time graphs 86, 87
distillation 38
disturbed habitats 33
division of cells 116, 117, 119
DMSA chemical 272
DNA (deoxyribonucleic acid) 37, 108, 114–116, 120–123, 125, 126
double helical structures 121

Earth 136–142, 146, 147, 150–155
eclipses 55
Ecstasy 205
effectors 201–203, 206
egg cells 114, 118, 119, 128
electrical power 182, 183
electricity 63–65, 140, 155, 156, 186–189
    circuits 164, 165
    ionic compounds 147
    motors 185
    nuclear power 275
electrodes 63, 154, 155
electrolysis 63–65, 154, 155
electrolytes 154, 155
electromagnetic induction 186
electromagnets 184
electrons 57–61, 148, 149, 167, 168, 170
elements 46–49, 137, 138, 146
    electrons 60, 61
    halogens 52, 53
    ionic theory 65
    radioactive materials 256
embryos 108, 126, 127
endothermic reactions 236, 237
energy 16, 58, 96–99, 274, 275, 278, 279
    body temperature 19
    chemical reactions 236, 237
    metal extraction 151
    photosynthesis 23
    respiration 34–37
environment 32, 33, 134, 135, 151, 159
enzymes 17–19, 20–23, 122, 124, 126
epilepsy 207
equations 50, 51, 90

Escherichia coli 196
ethanoic acid 242
ethanol 37, 38
ethics 211
excretion 16
exothermic reactions 236, 237
explosions 238
extraction of metals 150–152, 154, 159
eye 200, 202

Faraday, Michael 64
fatty axon sheaths 203
feldspar (mineral) 146
female gametes 114, 118, 119, 128
fermentation/fermenters 17, 38, 39
fetus 108
filament lamps 176
fine chemicals 226
fireworks 76
flame colour 54
flies 106, 197
fluorine 52, 53, 61
fluorite 52
food 16, 19
football 90
forces 88–90, 94, 95, 138, 166, 167, 278
 friction 80, 81
 magnets 184
 motion 76, 77
formula mass 140, 153, 248
formulae 51
freefall 83, 88
friction 80, 81, 94
fuel 38, 39
functional magnetic resonance imaging (fMRI) 207
fungi 37

gametes 114, 118, 119, 128
gamma radiation 257, 262, 263, 272
gases 24–27, 30, 51, 63, 68
 Earth 138
 fuel 39
 radioactive materials 267
 reaction rates 238, 239
generators 168, 186, 188, 189
genes 114, 115, 125–127
genetic switches 125–127
genetic variation 119
germanium 46
giant covalent structures 148, 149
giant ionic lattices 147
giant sequoia tree 109

glands 201
glucose 22, 23, 27
gold 146, 150
grafting procedures 128
graphite 148, 149
grasping reflexes 198
gravitational potential energy 97–99
gravity 77, 83, 88
greenhouses 29, 31
groups (periodic table) 47, 61
growth 16, 23, 104, 105
guanine (G) 120, 121

habitats 32, 33
Hadron Collider 57
haematite 146
hair cells 126
half-life 270, 271
halogens 52, 53, 67–69
harmful substances 52
hazards 265, 266, 276, 277
Health Protection Agency (HPA) 265, 266
heating 174
helium 54, 55, 278, 279
high-level waste (HLW) 276, 277
hormones 201
human simple reflexes 198–201
hydrogen 50, 51, 58, 138, 139, 232, 238, 278, 279
hydrosphere 136, 137, 140–142
hydroxides 144, 231, 232

imaging techniques 207, 211, 272
impulses 112, 113, 206
indicators 230
industry (chemical) 226, 227, 242
information 216, 217, 218
inheritance 120, 121
innate behaviour 198
insoluble carbohydrates 27
instantaneous speed 85
interactions 76–79
intermediate-level waste (ILW) 277
involuntary reflexes 196–201
iodide 145
iodine 52, 53, 61, 273
ionising radiation 256, 263, 267
ions 28, 64–67, 146, 147, 155
 chemical species 69
 salts 141–145, 232, 233
iron 152
iron sulfide 62
irradiation 263, 265

isotopes 262, 263, 266, 270–273
ITER nuclear fusion project 279

jet engines 78
joules (J) 97

kinetic energy 96, 98
knee jerk reflexes 198

lactic acid 36
laws of motion 94, 95
learning 208–211, 212, 213
leaves 25–27
leukaemia 129
life processes 14, 15
light intensity 29–31
light-dependent resistors (LDRs) 176
lightning 166
limestone 143, 146, 231, 234
limiting factors 31
line spectra 54, 55, 58
liquids 51
lithium 48, 49, 54, 61
lithosphere 136, 137, 146, 147, 150–155
living things 16, 17
lock-and-key models 18
long-term memory 214, 218
Low-Level Waste (LLW) 276
lubricants 149
lung cancer 267, 269

magnesium chloride 141
magnesium sulfate 244–247
magnetic resonance imaging (MRI) 207, 211
magnets 77, 184, 186, 187
mains electricity 188
male gametes 114, 118, 119
malleable metal properties 156
mantle of Earth 136
mass 248, 249
MDMA (Ecstasy) 205
medical imaging 207, 211, 272
meiosis 119
melted ionic solids 147
melting points 156
membranes 26, 34, 37
memory 210, 214–219
Mendel, Gregor 120, 121
Mendeleev, Dmitri 46
meristem cells 109, 111
messenger RNA (mRNA) 123
metals 46–49, 62–66, 141–147, 150–159
 acids 230, 231
 chemical species 69
 halogens 53

methane 39
mind 194, 195
minerals 28, 146, 148–150
mining 158
mitochondria 34, 37, 116
mitosis 116, 117
molecules 50, 52, 138, 139
molten salts 63
momentum 89–91, 94
motion 16, 74, 75, 94, 95
motor neurons 202, 203
motors 184, 185
movement 16, 74, 75
multistore memory models 217
muscles 201

National Grid 189
natural environment 134, 135
negative electric charge 166, 167
negative ions 65, 67
nerve cells 107, 203
nerve impulses 201–205
nerve pathways 212, 213
nerves 203
nervous system 201–203
neurons 202–205, 212, 13
neutralisation reactions 232, 233
neutrons 57, 260
newborn reflexes 198, 199
newts 110
nitrate ions 28
nitrogen 28, 137, 138, 140
non-metals
 chemical species 62–65, 67–69, 141–147
 Earth 138
 halogens 52, 53
 periodic table 47
Nuclear Decommissioning Agency (NDA) 276, 277
nuclear fission 274, 275
nuclear fusion 278, 279
nuclear power 274, 275
nuclear waste 276, 277
nuclear weapons 274, 275
nuclei 34, 56, 57, 114, 115
 plant cells 26
 radioactive materials 257–261, 270, 271, 274, 275, 278, 279
'Nun Study' (brain) 214
nutrition 16

Ohm's law 175, 176
orbits 58
ores 146, 150–152, 154, 158, 159
organelles 22, 26, 34, 37, 116

organs 107, 200–202, 206
ovaries 118
oxides 150, 231
oxygen 22, 34, 35, 50, 51, 137–140, 146
ozone layer 68

parallel circuits 172, 178, 181
pathways (neurons) 212, 213
Pavlov's experiments 208
Penfield, Wilder 207
percentage yield 249
periodic table 46–49, 52, 53, 60, 61
periods (periodic table) 47, 60
peripheral nervous system 203
pH 21, 230
phloem 26, 107
photosynthesis 16, 22, 23, 25, 26, 29–31
phototropism 112, 113
physical properties (alkali metals) 49
pigment 22
plant (chemical industry) 226
plants 16, 17, 22–36, 107, 109–113, 117
plasma 279
plastics 166
polymers 35
positive charge 166, 167, 278
positive ions 65, 66
potassium 48, 49, 61
potassium chloride 141
potential difference (p.d.) 179–181
potential dividers 180
potential energy 97–99
power, electrical 182, 183
power stations 188, 189, 274, 275, 278, 279
precipitates 143–145
products 50, 246, 247
proteins 18–23, 28, 122–127
proton numbers 57, 60
protons 57, 260
pupil reflexes 198
purity (chemical) 234, 235, 246, 247
pyrite 62, 146

quadrats 32, 33
quartz 146

radiation, chemical species 68
radiation dose 264, 265
radioactive decay 257, 270
radioactive materials 254, 255
radioiodine treatment 273
radiotherapy 273
radon 267–269, 271
random quadrat placement 33
rate of photosynthesis 29–31
reactants 50
reacting masses 248, 249
reaction rates 20, 21, 238–241
reaction of surfaces 82, 83
reactions (neutralisation) 232, 233
reactive metals 150, 154
receptors 200, 202, 203, 206
recycling 159
red blood cells 114
reduction reactions 150–152
reflex arcs 202
reflexes 196–201, 208, 209
rehearsal-based memory 218
relative atomic mass 46, 153
relative formula mass 153, 248
relay neurons 202
repetition-based learning 213, 218
reproduction 16, 117–119
repulsive forces 166, 167, 278
resistance 83, 173–177, 181
resistors 174, 176–178, 181
respiration 16, 23, 34–37, 126
responses to stimuli 16, 196
retrieval of information 216, 217
ribosomes 123
risk 93, 245, 265, 266, 269, 276, 277
RNA (ribonucleic acid) 123
rockets 78
rocks 137, 143, 146
root hair cells 28
rooting powder 111
rooting reflexes 199
rubbing (charge) 166
Rutherford, Ernest 258, 259

safety 91–93, 265, 266, 276, 277
salivary gland cells 126
salts 49, 62–65, 140–147, 230–233
samples 32, 33
sea water 141, 146
seat belts 92
sedimentary rocks 143
Sellafield (reprocessing site) 277
sense organs 200, 202, 206
senses 16
'sensory homunculus' 206
sensory memory storage 215
sensory neurons 202

sequoia tree 109
series circuits 171, 178, 180
sex cells 114, 118, 119, 128
sexual reproduction 118, 119
shells 58, 60, 61
short-term memory 214, 216
silica 146
silicon 146
silicon dioxide 149
simple behaviour 206
simple reflexes 196–201, 202
skin stem cells 128
slope 86
slowing down 95
small molecules 138
social behaviour 210
sodium 48, 49, 59, 61, 62, 64, 65, 69
sodium chloride 62, 64, 65, 69, 141, 142, 146, 147
sodium hydroxide 144
sodium sulfate 141
soil minerals 28
'solar system' models 258
solids 51
solubility 27, 142, 143
solution salts 63
specialised cells 107, 110, 124–127, 203
spectra 54, 55, 58
spectroscopy 54, 55
speed 84, 85, 98
speed–time graphs 87, 88
sperm cells 114, 118, 119
spheres of Earth 136–142, 146, 147, 150–155
starch molecules 23, 27
startle reflexes 199
state symbols 51
static electricity 167
steel 159
stem cells 108, 110, 128, 129
stepping reflexes 198
sterilisation (radioactive materials) 263
stimuli 16, 196, 200–204, 212, 213
stomata 26, 27, 30
strong metal properties 156
structural proteins 124
sucking reflexes 199
sulphates 145
Sun 55, 278
surface areas 240, 241, 243
surface reactions 82, 83
survival in nature 196, 197
sustainability 38
swimming reflexes 199
synapses 204, 205

tables 82, 83
tarnish (metals) 48

technetium-99m (radioactive tracer) 270–273
temperature 19–21, 29, 240, 241
testes 118
testing water 142
theoretical yield 249
therapeutic cloning 128
thermistors 176, 177
thymine (T) 120, 121
thyroid cancer 273
tissues 107
titrations 234, 235
toxic substances 52, 68, 205
transects (samples) 33
transferring energy 96
transformers 188, 189
treatment (medical) 272, 273
trends (chemical properties) 49

ultraviolet (UV) radiation 68
undisturbed habitats 33
unspecialised cells 110
unstable nuclei 257
uracil (U) 123

vacuoles 26
van de Graaff generators 168
variable resistors 176, 177
velocity 85
  see also speed
voltage 173, 179, 186–189
voltmeters 179

walking palm tree 112
walls 83
waste 16, 39, 276, 277
water 30, 48, 137, 138, 140–143, 147
  chemical equations 50, 51
  diffusion 24, 25
watts (W) 182
weather 141, 166
work 96–99, 183
working memory 218, 19

xylem 26, 107

yeast cells 37, 38, 117
yield 29, 247, 249

zinc 152
zinc chloride 63
zinc ions 144
zygotes 108, 118, 119

Appendices

Useful relationships, units, and data
Relationships
You will need to be able to carry out calculations using these mathematical relationships.

P4 Explaining motion
$$\text{speed} = \frac{\text{distance}}{\text{time}}$$

$$\text{acceleration} = \frac{\text{change in velocity}}{\text{time taken}}$$

momentum = mass × velocity

change of momentum = resultant force × time for which it acts

work done by a force = force × distance moved in the direction of the force

amount of energy transferred = work done

change in gravitational potential energy = weight × vertical height difference

kinetic energy = ½ × mass × velocity²

P5 Electric circuits
power = voltage × current

$$\text{resistance} = \frac{\text{voltage}}{\text{current}}$$

$$\frac{\text{voltage across primary coil}}{\text{voltage across secondary coil}} = \frac{\text{number of turns in primary coil}}{\text{number of turns in secondary coil}}$$

C6 Chemical synthesis
$$\text{percentage yield} = \frac{\text{actual yield}}{\text{theoretical yield}} \times 100\%$$

P6 Radioactive materials
Einstein's equation: $E = mc^2$, where E is the energy produced, m is the mass lost, and c is the speed of light in a vacuum.

Units that might be used in the Additional Science course

length: metres (m), kilometres (km), centimetres (cm), millimetres (mm), micrometres (μm), nanometres (nm)

mass: kilograms (kg), grams (g), milligrams (mg)

time: seconds (s), milliseconds (ms)

temperature: degrees Celsius (°C)

area: cm^2, m^2

volume: cm^3, dm^3, m^3, litres (l), millilitres (ml)

speed and velocity: m/s, km/s, km/h

energy: joules (J), kilojoules (kJ), megajoules (MJ), kilowatt-hours (kWh), megawatt-hours (MWh)

electric current : amperes (A), milliamperes (mA)

potential difference/voltage: volts (V)

resistance: ohms (Ω)

power: watts (W), kilowatts (kW), megawatts (MW)

radiation dose: sieverts (Sv)

Prefixes for units

nano	micro	milli	kilo	mega	giga	tera
one thousand millionth	one millionth	one thousandth	× thousand	× million	× thousand million	× million million
0.000000001	0.000001	0.001	1000	1000 000	1000 000 000	1000 000 000 000

Useful information and data

P4 Explaining motion
A mass of 1 kg has a weight of 10 N on the surface of the Earth.

C5 Chemicals of the natural environment
Approximate proportions of the main gases in the atmosphere: 78% nitrogen, 21% oxygen, 1% argon, and 0.04 % carbon dioxide.

P5 Electric circuits
mains supply voltage: 230 V

P6 Radioactive materials
speed of light (c) = 300 000 000 m/s

Chemical formulae

C4 Chemical patterns
water H_2O, hydrogen H_2, chlorine Cl_2, bromine Br_2, iodine I_2

lithium chloride LiCl, lithium bromide LiBr, lithium iodide LiI

sodium chloride NaCl, sodium bromide NaCl, sodium iodide NaI

potassium chloride KCl, potassium bromide KBr, potassium iodide KI

lithium hydroxide LiOH, sodium hydroxide NaOH, potassium hydroxide KOH

C5 Chemicals of the natural environment
nitrogen N_2, oxygen O_2, argon A, carbon dioxide CO_2

sodium chloride NaCl, magnesium chloride $MgCl_2$

sodium sulfate Na_2SO_4, magnesium sulfate $MgSO_4$

potassium chloride KCl, potassium bromide KBr

C6 Chemical synthesis
chlorine Cl_2, hydrogen H_2, nitrogen N_2, oxygen O_2

hydrochloric acid HCl, nitric acid HNO_3, sulfuric acid H_2SO_4

sodium hydroxide NaOH, sodium chloride NaCl, sodium carbonate Na_2CO_3, sodium nitrate $NaNO_3$, sodium sulfate Na_2SO_4, potassium chloride KCl

magnesium oxide MgO, magnesium hydroxide $Mg(OH)_2$, magnesium carbonate $MgCO_3$, magnesium chloride $MgCl_2$, magnesium sulfate $MgSO_4$

calcium carbonate $CaCO_3$, calcium chloride $CaCl_2$, calcium sulfate $CaSO_4$

Qualitative analysis data

Tests for negatively charged ions

Ion	Test	Observation
carbonate CO_3^{2-}	add dilute acid	effervesces, and carbon dioxide gas is produced (the gas turns lime water milky)
chloride (in solution) Cl^-	acidify with dilute nitric acid, then add silver nitrate solution	white precipitate
bromide (in solution) Br^-	acidify with dilute nitric acid, then add silver nitrate solution	cream precipitate
iodide (in solution) I^-	acidify with dilute nitric acid, then add silver nitrate solution	yellow precipitate
sulfate (in solution) SO_4^{2-}	acidify, then add barium chloride solution or barium nitrate solution	white precipitate

Tests for positively charged ions

Ion	Test	Observation
calcium Ca^{2+}	add sodium hydroxide solution	white precipitate (insoluble in excess)
copper Cu^{2+}	add sodium hydroxide solution	light-blue precipitate (insoluble in excess)
iron(II) Fe^{2+}	add sodium hydroxide solution	green precipitate (insoluble in excess)
iron(III) Fe^{3+}	add sodium hydroxide solution	red–brown precipitate (insoluble in excess)
zinc Zn^{2+}	add sodium hydroxide solution	white precipitate (soluble in excess, giving a colourless solution)

OXFORD
UNIVERSITY PRESS

Great Clarendon Street, Oxford OX2 6DP

Oxford University Press is a department of the University of Oxford.
It furthers the University's objective of excellence in research,
scholarship, and education by publishing worldwide in

Oxford   New York

Auckland   Cape Town   Dar es Salaam   Hong Kong   Karachi
Kuala Lumpur   Madrid   Melbourne   Mexico City   Nairobi
New Delhi   Shanghai   Taipei   Toronto

With offices in
Argentina   Austria   Brazil   Chile   Czech Republic   France   Greece
Guatemala   Hungary   Italy   Japan   Poland   Portugal   Singapore
South Korea   Switzerland   Thailand   Turkey   Ukraine   Vietnam

Oxford is a registered trade mark of Oxford University Press
in the UK and in certain other countries.

© University of York and the Nuffield Foundation 2011.

The moral rights of the authors have been asserted.

Database right Oxford University Press (maker).

First published 2011.

All rights reserved. No part of this publication may be reproduced,
stored in a retrieval system, or transmitted, in any form or by any means,
without the prior permission in writing of Oxford University Press, or as
expressly permitted by law, or under terms agreed with the appropriate
reprographics rights organization. Enquiries concerning reproduction
outside the scope of the above should be sent to the Rights Department,
Oxford University Press, at the address above.

You must not circulate this book in any other binding or cover
and you must impose this same condition on any acquirer.

British Library Cataloguing in Publication Data.

Data available.

ISBN 978-0-19-913820-3

10 9 8 7 6 5 4 3 2 1

Printed in Great Britain by Bell and Bain, Glasgow.

Paper used in the production of this book is a natural, recyclable product
made from wood grown in sustainable forests. The manufacturing process
conforms to the environmental regulations of the country of origin.

Acknowledgements

The publisher and authors would like to thank the following for their permission to reproduce photographs and other copyright material:
P13t: David Taylor/Science Photo Library; **P13b:** John Howard/Science Photo Library; **P14:** Addimage/Istockphoto; **P17t:** Tfoxfoto/Shutterstock; **P17b:** James King-Holmes/Celltech R&D Ltd/Science Photo Library; **P18:** J.C. Revy, Ism/Science Photo Library; **P19:** Eric And David Hosking/Corbis; **P20:** Simon Fraser/Science Photo Library; **P22:** J.C. Revy, ISM/Science Photo Library; **P23t:** Biophoto Associates/Science Photo Library; **P23b:** Dr Jeremy Burgess/Science Photo Library; **P24:** Zooid Pictures; **P25:** Corbis; **P27:** Biophoto Associates/Science Photo Library; **P29:** Sinclair Stammers/Science Photo Library; **P30t:** Dr Jeremy Burgess/Science Photo Library; **P30b:** Dr Jeremy Burgess/Science Photo Library; **P32:** P Phillips/Shutterstock; **P33:** Martyn F. Chillmaid/Science Photo Library; **P34:** Kimimasa Mayama/Reuters; **P35:** K.R. Porter/Science Photo Library; **P36l:** flo Foto Agency/Alamy; **P36r:** Blickwinkel/Alamy; **P37:** Power And Syred/Science Photo Library; **P38t:** Viktor1/Shutterstock; **P38:** AZP Worldwide/Shutterstock; **P39t:** Eye of Science/Science Photo Library; **P39b:** Ashley Cooper/Specialiststock/Splashdowndirect/Rex Features; **P44:** Dirk Wiersma/Science Photo Library; **P46:** The Print Collector/Photolibrary; **P48:** Andrew Lambert Photography/Science Photo Library; **P49:** Charles D. Winters/Science Photo Library; **P51:** Don Grall, Visuals Unlimited/Science Photo Library; **P52:** Claude Nuridsany & Marie Perennou/Science Photo Library; **P53:** Andrew Lambert Photography/Science Photo Library; **P54:** David Taylor /Science Photo Library; **P55l:** Dept. Of Physics, Imperial College/Science Photo Library; **P55r:** Roger Ressmeyer/Corbis; **P57:** David Parker/Science Photo Library; **P58:** Dept. Of Physics, Imperial College/Science Photo Library; **P62tl-r:** Arnold Fisher/Science Photo Library, Andrew Lambert Photography/Science Photo Library, Charles D. Winters/Science Photo Library, Maximilian Stock Ltd/Science Photo Library; **P62b:** Dirk Wiersma/Science Photo Library; **P63t:** Andrew Lambert Photography/Science Photo Library; **P63b:** Charles D. Winters/Science Photo Library; **P68t:** Andrew Lambert Photography/Science Photo Library; **P68b:** NASA/Science Photo Library; **P69t:** Andrew Lambert Photography/Science Photo Library; **P69b:** Arnold Fisher/Science Photo Library; **P72:** Roger Ressmeyer/Corbis; **P74:** Paul Mckeown/Istockphoto; **P76:** Vakhrushev Pavel/Shutterstock; **P77:** Photodisc/Photolibrary; **P78:** NASA; **P79:** Greg McCracken/Istockphoto; **P81:** Power And Syred/Science Photo Library; **P83:** Kenneth Eward/Biografx/Science Photo Library; **P85:** Santiago Cornejo/Shutterstock; **P88:** Kenneth Eward/Biografx/Science Photo Library; **P90:** Offside/Rex Features; **P91:** Corbis; **P102:** American Institute Of Physics/Science Photo Library; **P104:** Neil Bromhall/Science Photo Library; **P106tl:** Oxford Scientific Films/Photolibrary; **P106tm:** Oxford Scientific Films/Photolibrary; **P106tr:** Corel/Oxford University Press; **P106bl:** Michael & Patricia Fogden/Corbis; **P106bm:** Corel/Oxford University Press; **P106br:** Corel/Oxford University Press; **P108t:** Alexander Tsiaras/Science Photo Library; **P108b:** Science Photo Library; **P109:** Bob Rowan/Progressive Image/Corbis; **P134:** Dane Steffes/Istockphoto; **P139t-b:** Adam Hart-Davis/Science Photo Library, Nina Towndrow/Nuffield Curriculum Centre, Nina Towndrow/Nuffield Curriculum Centre, Adam Hart-Davis/Science Photo Library; **P140:** Galen Rowell/Corbis; **P142l:** Joe Gough/Shutterstock; **P142m:** Martin Garnham/Dreamstime; **P142r:** Dogi/Shutterstock; **P143l:** Kesu/Shutterstock; **P143r:** Charles D. Winters/Science Photo Library; **P144l:** Martyn F. Chillmaid/Science Photo Library; **P144r:** Andrew Lambert Photography/Science Photo Library; **P145t:** Andrew Lambert Photography/Science Photo Library; **P145b:** Martyn F. Chillmaid/Science Photo Library; **P146t:** Jose Manuel Sanchis Calvete/Corbis; **P146m:** George Bernard/Science Photo Library; **P146b:** Arnold Fisher/Science Photo Library; **P148:** Mikeuk/Istockphoto; **P149:** Sinclair Stammers/Science Photo Library; **P150:** Layne Kennedy/Corbis; **P151:** James L. Amos/Corbis; **P152:** Kevin Fleming/Corbis; **P154:** Maximilian Stock Ltd/Science Photo Library; **P155:** Uranium; **P156tl:** H. David Seawell/Corbis; **P156tm:** Alexis Rosenfeld/Science Photo Library; **P156tr:** John Van Hasselt/Corbis; **P156b:** Charles E. Rotkin/Corbis; **P158:** Nik Wheeler/Corbis; **P162t:** Dirk Wiersma/Science Photo Library; **P162b:** H. David Seawell/Corbis; **P164:** Blackred/Istockphoto; **P166t:** OAR/ERL/National Severe Storms Laboratory (NSSL) NOAA Photo Library; **P166b:** Charles D. Winters/Science Photo Library; **P172:** Anthony Redpath/Corbis; **P173:** Zooid Pictures; **P176:** Masterfile; **P178:** Ilya Zlatyev/Shutterstock; **P180:** Lisa F. Young/Shutterstock; **P183:** Philippe Hays/Rex Features; **P185:** Okea/Istockphoto; **P187:** Colin Cuthbert/Science Photo Library; **P188t:** Anthony Vizard/Eye Ubiquitous/Corbis; **P188b:** Scottish Power; **P189l:** British Energy; **P189m:** Nicholas Bailey/Rex Features; **P189r:** Eye of Science/Science Photo Library; **P192:** Ilya Zlatyev/Shutterstock; **P194:** Sovereign, Ism/Science Photo Library; **P196t:** Dennis Kunkel/phototake Inc./Alamy; **P196bl:** Astrid & Hanns-Frieder Michler/Science Photo Library; **P196br:** Eye of Science/Science Photo Library; **P197tl:** Jeff Rotman/Nature Picture Library; **P197tr:** Tobias Bernhard/Oxford Scientific Films/Photolibrary; **P197bl:** Manfred Danegger/NHPA; **P197br:** Clive Druett/Papilio/Corbis; **P198l:** Sheila Terry/Science Photo Library; **P198m:** Owen Franken/Corbis; **P198r:** Laura Dwight/Corbis; **199l:** BSIP Astier/Science photo Library; **P199r:** Jennie Woodcock/Reflections Photolibrary/Corbis; **P199b:** Morgan McCauley/Corbis; **P202:** Adam Hart-Davis/Science Photo Library; **P204:** Greg Fiume/Corbis; **P205:** Sipa Press/Rex Features; ; **P207t:** S.Kramer/Custom Medical Stock Photo/Science Photo Library; **P207b:** Mark Lythgoe and Steve Smith/Wellcome Trust; **P208:** Steve Bloom Images/Alamy; **P210t:** Lawrence Manning/Corbis; **P210bl:** Sally and Richard Greenhill/Alamy; **P210br:** Corbis; **P213:** Jerry Wachter/Science Photo Library; **P214:** Karen Kasmauski/Corbis; **P215:** Roger Ressmeyer/Corbis; **P218:** c.Warner Br/Everett/Rex Features; **P222:** Anthony Bannister/Gallo Images/Corbis; **P224:** Michael Rosenfeld/Science Faction/Corbis; **P226:** Maximilian Stock Ltd/Science Photo Library; **P227t:** Geoff Tompkinson/Science Photo Library; **P227b:** William Taufic/Corbis; **P228l:** XXX; **P228r:** Martyn F. Chillmaid/Science Photo Library; **P229l:** Martyn F. Chillmaid/Science Photo Library; **P229r:** Lurgi Metallurgic/Outokumpu; **P230l:** Charles D. Winters/Science Photo Library; **P230m:** Marketa Mark/Shutterstock; **P230l:** Kojiro/Shutterstock; **P235:** Martyn F. Chillmaid; **P236l:** Martyn F. Chillmaid/Science Photo Library; **P236m:** John Casey/Fotolia; **P236r:** AJ Photo/Science Photo Library; **P237:** Pablo Bartholomew/Getty Images News/Getty Images; **P238:** Peter Bowater/Alamy; **P240:** Life Hacker; **P242:** Gary Banks/BP Saltend; **P244:** Holt Studios International; **P247:** Sidney Moulds/Science Photo Library; **P252:** Peter Bowater/Alamy; **P254:** EFDA-JET/Science Photo Library; **P256t:** Radiation Protection Division/Health Protection Agency/Science Photo Library; **P256bl:** Health Protection Agency/Science Photo Library; **P256br:** Kletr/Shutterstock; **P257:** Davies and Starr/Stone/Getty Images; **P260t:** Amrit G/Fotolia; **P260:** Istockphoto; **P263t:** Shutterstock; **P263b:** Geoff Tompkinson/Science Photo Library; **P265:** Science Photo Library; **P266:** Josh Sher/Science Photo Library; **P266b:** Health Protection Agency/Science Photo Library; **P267:** Health Protection Agency/Science Photo Library; **P272:** Prof. J. Leveille/Science Photo Library/Science Photo Library; **P274:** Rex Features; **P275t:** Science Photo

Library; **P275b:** Ria Novosti/Science Photo Library; **P276t:** Steve Allen/Science Photo Library; **P276b:** Science Photo Library; **P278:** European Space Agency/Science Photo Library; **P282:** Geoff Tompkinson/Science Photo Library.

Illustrations by IFA Design, Plymouth, UK, Clive Goodyer, and Q2A Media.
The publisher and authors are grateful for permission to reprint the following copyright material:
Although we have made every effort to trace and contact all copyright holders before publication this has not been possible in all cases. If notified, the publisher will rectify any errors or omissions at the earliest opportunity.

Project Team acknowledgements
These resources have been developed to support teachers and students undertaking the OCR GCSE Science Twenty First Century Science suite of specifications. They have been developed from the 2006 edition of the resources.
We would like to thank David Curnow and Alistair Moore and the examining team at OCR, who produced the specifications for the Twenty First Century Science course.

Mixed Sources
Product group from well-managed forests and other controlled sources
www.fsc.org Cert no. TT-COC-002769
© 1996 Forest Stewardship Council

Authors and editors of the first edition
We thank the authors and editors of the first edition, Jenifer Burden, Simon Carson, John Holman, Anna Grayson, Angela Hall, John Holman, Andrew Hunt, Bill Indge, John Lazonby, Allan Mann, Jean Martin, Robin Millar, Nick Owens, Stephen Pople, Carol Tear, and Linn Winspear.
Many people from schools, colleges, universities, industry and the professions contributed to the production of the first edition of these resources. We also acknowledge the invaluable contribution of the teachers and students in the Pilot Centres.
The first edition of Twenty First Century Science was developed with support from the Nuffield Foundation, The Salters Institute, and The Wellcome Trust. A full list of contributors can be found in the Teacher and Technician resources.

The continued development of *Twenty First Century Science* is made possible by generous support from:
- The Nuffield Foundation
- The Salters' Institute